Design For Inclusion

Volume 6

Series Editor

Giuseppe Di Bucchianico, Department of Engineering and Geology, University of Chieti-Pescara, Pescara, Italy

The book series "Design for Inclusion" publishes studies aimed at untangling the complex relationships between design and society, especially with regard to human diversity and inclusion. The series cover the classical fields of industrial design, such as product design, interior design, service design and communication design, with a special focus on their interactions with other disciplines, such as the physical and social sciences. In particular, the series reports on the latest developments, both theoretical and practical, concerning different approaches to "Design for Inclusion": Universal Design, Inclusive Design, and Design for All. It shares insights into their multiple, interdisciplinary relationships with both the human sciences and the transversal frameworks and approaches typical of Human-Centered Design (HCD) and Design for Sustainability. "Design for Inclusion" addresses a very broad readership, including designers, engineers, architects, social scientists and stakeholders, who deal with any of the topics mentioned above and their various implications. It publishes volumes fostering scientific advances, as well as books devoted to supporting education and professional training. These include monographs, edited books and conference proceeding. Outstanding PhD theses on emerging topics, if properly reworked, may also be considered for publication. This book series is published with the support of EIDD—Design for All Europe.

Stefania Camplone · Claudia Porfirione
Editors

Design for All and the Future of the Planet

Proceedings of EIDD 2024 Annual Conference, June 6–7, 2024, Vila Nova de Gaia, Portugal

EIDD DESIGN FOR ALL EUROPE

Springer

Editors
Stefania Camplone
INGEO Department
University of Chieti-Pescara
Pescara, Italy

Claudia Porfirione
Architecture and Design Department
University of Genoa
Genova, Italy

ISSN 2662-5644 ISSN 2662-5652 (electronic)
Design For Inclusion
ISBN 978-3-032-21768-4 ISBN 978-3-032-21769-1 (eBook)
https://doi.org/10.1007/978-3-032-21769-1

This work was supported by EIDD—Design for All Europe.

This Springer imprint is published by the registered company Springer Nature Switzerland AG
The registered company address is: Gewerbestrasse 11, 6330 Cham, Switzerland

Foreword

On June 6–7, 2024, the international conference *"Gaia 2024. Design for All and the Future of the Planet"* was held in Vila Nova de Gaia, Portugal, a city located on the south bank of the Douro River, facing Porto. The Conference was organized by the international association EIDD—Design for All Europe, which I have the honor of chairing. For over a year, EIDD has been engaged in deep reflection on several emerging global challenges facing contemporary societies, such as human diversity and aging, conflict and humanitarian emergencies, climate change, and the development of new technologies, that demand a conscious and responsible response.

This process culminated in the drafting of a new document which, 20 years after the foundational *Stockholm Declaration* of 2004, reaffirms and updates its principles in response to the pressing issues now facing humanity.

The International Conference *"Gaia 2024. Design for All and the Future of the Planet"* provided a valuable opportunity to bring together experts—professionals, researchers, developers, policymakers, and users—to share knowledge, discuss progress, develop new ideas, and engage in dialogue around the outcomes of academic research, applied experiences, and practical achievements related to Design for All. The presentations delivered during the academic sessions and discussion panels were introduced by distinguished keynote speakers, whom I would like to personally thank for their valuable contributions to the collective reflection: Rama Gheerawo, Director of the Helen Hamlyn Centre for Design, Royal College of Art, on *Artificial Intelligence;* Rafael Mariano Grossi, Director General at the International Atomic Energy Agency, UN, on *Energy*; Pete Kercher, EIDD Ambassador, on *Climate Change*; Gennady Trukhanov, Mayor of the City of Odessa, on *Conflicts and Emergencies.*

At the same time, the Conference also provided the ideal stage for the official presentation of the *Gaia Declaration.* This new document was the result of a collaborative effort involving all EIDD—Design for All Europe members, united in the goal of addressing the major social, economic, environmental, and cultural challenges of our time, in line with the core values and approach of Design for All. The name "Gaia" given to the Declaration is the product of a fortunate coincidence: it refers both to the city that hosted the event, *Vila Nova de Gaia*, and to the ancient Greek goddess Gaia, who gave her name to our planet Earth.

The event in Vila Nova de Gaia thus represents the outcome of a collective commitment by the EIDD Member Organizations, as well as the dedication of a small group of colleagues and friends who helped organize it. Their names are listed in the Conference Organization list, but I would like to mention them personally: Pete, Nuno, Markus, and Maria Graça.

This publication, which includes the Conference Proceedings along with a summary of the Gaia Declaration Ceremony, is especially valuable.

Unlike other approaches to inclusive design that have had more time to build a solid body of publications, Design for All still has a relatively young and underrepresented literature. For this reason, this volume contributes meaningfully to closing that gap, especially since it is published as an *Open Access* resource, freely available to all.

Moreover, this book is particularly important for EIDD—Design for All Europe, as it represents the first official publication directly resulting from the Association's work, supported by a Scientific Committee, which was responsible for the selection and peer review of the contributions.

An additional source of pride is that this book is part of the *"Design for Inclusion"* series published by the prestigious international publisher Springer, of which EIDD is an official supporter.

I would like to sincerely thank the conference partners, that are the City Council of Vila Nova de Gaia, the Tourism Authority of Porto and Fundación ONCE, for their practical and economical contribution to the realization of the Conference and of this book of proceedings; and I would also like to thank all the authors who chose the *Gaia 2024 International Conference* as the venue to present their work. We have all appreciated their commitment, and readers will surely recognize the quality of the more than twenty contributions included in this volume.

Finally, my warm thanks go to the editors of this publication, Claudia and Stefania, for the care and attention with which they have carried out their work.

It is, therefore, with great pleasure and pride that I present to you the Proceedings of the International Conference *"Gaia 2024. Design for All and the Future of the Planet."*

Vila Nova De Gaia, Portugal

Giuseppe Di Bucchianico
President of EIDD – Design for All Europe

Preface

Twenty years ago, the Stockholm Declaration defined Design for All as "design for human diversity, social inclusion and equality." Its goal is to improve the quality of life for all individuals by valuing their specific characteristics, through a holistic approach to design processes and methods, is based on the conscious and active participation of all stakeholders involved in the design chain, rather than on a simple exchange between designer and client. This includes economic and political decision-makers, designers and manufacturers, and ultimately, the end users of spaces, products, services and systems, all of whom share the common goal of enabling *independent* use by people with diverse needs and abilities.

Similarly, a Design for All approach brings "cross-cutting" benefits that go beyond individual design actions, impacting both the social and economic spheres at different scales and over time. Whether applied at the level of a single company or a territorial authority, the key principles and methods remain consistent. Operationally, the goal is to pursue design solutions that are immediately usable by the majority of people without the need for specific modifications or, when necessary, that can be easily adapted to meet the physical, sensory, or cognitive abilities of different user profiles.

It has often been pointed out that one of the most innovative aspects of Design for All lies in its shift from a focus on mere "use" to the broader concept of "experience": this transition, from "user" to "experiencer," implies not only meeting a need but also responding to aspirations and desires. In other words, the aim is not only to ensure a project's accessibility and usability (which remain essential conditions) but also to promote the well-being of all, taking into account a broader range of human characteristics and assessing them holistically and collaboratively. This inclusive design approach extends active, comfortable, and enjoyable use of environments, products, and services to a wide variety of individuals with different perceptual, motor, and cognitive abilities, thus becoming a powerful driver of competitive renewal, precisely because it seeks to engage user groups that have previously been excluded from access to goods and services.

The first European organization to systematically and collaboratively address Design for All was the **European Institute for Design and Disability (EIDD)**. Founded in Dublin in 1993 under the leadership of Paul Hogan, EIDD quickly became a reference point, a space for debate and initiative aimed at concretely overcoming human–environment conflicts through design. The Institute soon shifted its focus from design for disability to "Design for All," sparking a dialogue that eventually led to the development of a specific design approach, which was formalized in the 2004 Stockholm Declaration, whose principles were later recognized by the European Union.

Twenty years after the Stockholm Declaration, Design for All continues to gain increasing attention across Europe and globally, establishing itself as a field with its own scientific community that shares and upholds its values. This growing recognition is undoubtedly a positive sign.

At the same time, however, it is essential to respond to the social, political, economic, and technological changes that are reshaping today's world. In many ways, contemporary societies are profoundly different from those of two decades ago. Demographic shifts, environmental and social emergencies, and the unstoppable and rapidly accelerating pace of technological development call for a critical reflection on the very principles that underpin the Stockholm Declaration.

The annual gathering of the Member Organizations of **EIDD—Design for All Europe**, held on 2024 in Vila Nova de Gaia (Portugal), provided the opportunity to organize the International Conference *"Gaia 2024—Design for All and the Future of the Planet."* The event served both as a moment of reflection on what has been accomplished and what continues to be done in academic research and applied practice relating to Design for All, and as a moment of relaunch, aimed at addressing emerging issues that the Design for All community must inevitably confront in the near future.

The Conference was structured around a series of academic sessions and discussion panels, where research findings and applied Design for All experiences were presented, mainly within the domains of *"Spaces and Environments"* and *"Products and Services."* These sessions were interspersed with keynote speeches from internationally renowned figures, addressing four pressing themes considered crucial for the future development of Design for All:

Rama Gheerawo, Director of the Helen Hamlyn Centre for Design (London), addressed the topic of Artificial Intelligence;

Rafael Mariano Grossi, Director General of the UN International Atomic Energy Agency (Vienna), reflected on the theme of Energy;

Gennady Trukhanov, Mayor of the City of Odessa, highlighted the challenges posed by Conflict and Humanitarian Emergencies;

Pete Kercher, EIDD Ambassador, initiated a discussion on Climate Change and its possible consequences for economic and social systems.

The event in Vila Nova de Gaia culminated in the **Gaia Declaration Ceremony**, during which the *Gaia Declaration* was officially presented as a natural evolution and reinterpretation of the Stockholm Declaration's principles, revisited in light of the new and emerging challenges facing contemporary societies.

Designing for Inclusion: Interdisciplinary Perspectives on Spaces, Products, and Services

This volume brings together the proceedings of a multidisciplinary Conference centered around the theme of inclusive design, with particular attention to the evolving intersections between architecture, urban planning, health, education, and digital innovation. The event was structured into two primary parallel parts: "Spaces and Environments" and "Products and Services," reflecting a holistic approach to inclusion that transcends traditional disciplinary boundaries.

Seventeen contributions, presented by scholars and practitioners from across Europe and beyond, compose this compendium. Their diversity in methodology, scope, and geographic origin offers a rich panorama of how inclusive design principles are being interpreted, applied, and reimagined across contexts. Despite the variety, certain cross-cutting themes emerge clearly: the centrality of user engagement, the role of co-design and participatory methods, the integration of emerging technologies, and the increasing relevance of ecological and systemic frameworks in approaching inclusion.

The first part, "Spaces and Environments," emphasizes the spatial dimension of inclusivity, exploring how the built environment can promote equity, accessibility, and well-being for all individuals, regardless of ability, age, or circumstance.

K. Görner and colleagues' contribution, *Cities for All*, addresses Universal Design at the urban scale, proposing principles for incorporating inclusive thinking into spatial planning documentation. This macro-level vision is echoed in the paper by S. Majcher and L. Selcová, who advocate for embedding Universal Design as a core architectural expression rather than an afterthought. Similarly, M. Bezáková et al. examine how Slovakia's aging housing stock—specifically prefabricated buildings—can be adapted to meet the needs of older residents through flexible, inclusive solutions that prioritize both accessibility and comfort.

From urban planning to interior navigation, several papers highlight the importance of inclusive wayfinding. M. Więckowska and P. Rudnicka explore how virtual reality can replicate and study wayfinding behavior, opening new possibilities for accessible design evaluation. The Polish case study presented by A. Nawrocka and colleagues focuses on the implementation of inclusive signage systems, emphasizing co-creation, user testing, and interdisciplinarity.

Another recurring theme is the use of design to support vulnerable populations through spatial environments that emphasize dignity and community. *The Carpe Diem Dementia Village* in Norway, presented by C. Heier Anglero, is a powerful example of architecture that supports autonomy and inclusion for individuals with cognitive decline. Through thoughtful design and community integration, the project transcends the institutional paradigm of eldercare.

Zuzana Ceresnova and her team extend the conversation on adaptability in housing by showcasing national projects aimed at developing adaptable housing models aligned with universal design principles. Their work illustrates how policy, participatory planning, and design research can converge to address demographic challenges.

Another significant contribution is Janice Rieger's global case study on mapping Design for All (DfA) through ecological frameworks. Her paper emphasizes systems thinking and the co-creation of inclusive ecologies, shifting the focus from isolated design solutions to broader cultural and spatial ecosystems. This systems-based, ecological perspective is an emergent theme across several contributions, reflecting a growing awareness that inclusive design must operate at multiple interconnected levels.

Finally, the papers by E. Ballegaard et al. and A. Vacanti address the foundational role of research methodology in informing inclusive spatial design. Ballegaard's team analyzes how methodological choices influence the knowledge produced about user experience, while Vacanti proposes the "User Integration Funnel," a conceptual model that rethinks user engagement in human-centered and adaptive design processes. Together, these contributions form a compelling narrative on the transformation of physical environments through inclusive, participatory, and data-informed practices, while also posing critical questions about the balance between seamlessness and user agency.

The second part, "Products and Services," broadens the scope from architectural and environmental design to interactive systems, objects, and experiences. The presented works underscore the importance of inclusive practices in domains such as healthcare, education, cultural heritage, and consumer products.

Mental health is a central concern in the paper *Open your mind* by X. Ferrari Tumay and F. Lorusso, which details a participatory project involving individuals in mental health care through creative and digital design workshops. Their approach—based on co-design, empowerment, and serious leisure—exemplifies how inclusive service design can foster emotional well-being and social integration.

Education and learning environments are the focus of the paper by I. Poultourtzidis and colleagues, who present research on accessible educational escape rooms. Their work demonstrates how participatory and immersive methods can promote engagement and knowledge retention for students with disabilities, contributing to inclusive pedagogical strategies. The theme of multisensory learning is further developed by F. Tosi et al., who discuss how inclusive itineraries in Florentine museums enhance user engagement through tactile, auditory, and olfactory stimuli, particularly benefiting visitors with visual impairments.

Similarly, I. Leggiero and her team tackle the theme of cultural participation from the perspective of aging populations. Their research explores how digital tools, co-designed with older adults, can bridge the gap between cultural engagement and physical limitations, promoting active aging and well-being. The emphasis on co-creation and digital inclusion aligns this contribution with a broader movement toward design that empowers marginalized users through technology.

In the sphere of consumer products, F. Rodighiero introduces *Prime* and *Shine*, two inclusive washbasins designed to merge accessibility with aesthetic refinement. This paper underscores the importance of dignity and beauty in functional product design, suggesting that inclusivity need not compromise elegance.

Cultural sensitivity and intercultural competence are addressed innovatively through *O'Culo*, a serious game developed by K. Vaes and D. Van Boxem. By fostering dialogue and awareness around cultural differences in design, the game positions play as a powerful tool for reflexivity and empathy in design education and practice. It highlights an increasingly relevant area of inquiry: the intersection between inclusivity and global cultural diversity.

This book gathers the proceedings of the conference. The papers underwent a single-blind peer review process, in which each submitted paper was evaluated by one anonymous reviewer. The reviewers were selected among members of the scientific committee and academic experts in the relevant fields. A double evaluation was carried out, including both a pre-review phase and a post-review phase following the reviewers' comments.

Each paper was assessed according to the following criteria:

- Scientific relevance of the addressed subject
- Innovation, originality, and impact of the article
- Methodological rigor and coherence of the argumentation
- Quality and clarity of the structure
- Overall assessment aimed at improving the quality of the contribution

The seventeen accepted papers represent eight different nationalities: Slovakia, Denmark, Poland, Italy, Belgium, Greece, Norway, and Australia.

While the papers span diverse domains—from healthcare architecture to game design—they converge on several significant themes:

- User Participation and Co-Design: Nearly every contribution foregrounds the role of users, not just as beneficiaries but as active co-creators. Whether through workshops, surveys, or participatory simulations, these studies emphasize that inclusivity cannot be achieved without meaningful engagement.
- Multisensory and Emotional Design: Across environments and interfaces, authors stress the role of multisensory engagement in creating emotionally resonant and inclusive experiences. From tactile museum installations to beautiful and functional bathroom furnishings, sensory richness is framed as central to user well-being.
- Ecological and Systemic Approaches: Several contributions advocate for moving beyond isolated design solutions to consider systems and ecologies, whether social, technological, or environmental. This shift suggests a growing awareness of the complexity and interconnectedness of inclusive design challenges.
- The Role of Technology: VR, digital tools, and serious games appear across papers as both research instruments and mediums of inclusion.

Technology is positioned not merely as a facilitator, but as a domain to be shaped by inclusive values.

- Bridging Disciplines and Sectors: The proceedings reveal numerous instances of collaboration across academia, public institutions, industry, and civil society. Such partnerships are essential to developing inclusive practices that are contextually grounded and socially impactful.

In conclusion, this collection of proceedings reflects a vibrant, evolving field that recognizes inclusion as a multidimensional and shared responsibility. From cities and cultural institutions to educational tools and daily-use products, the contributions underscore that inclusive design is not a niche concern, but a comprehensive framework for rethinking how we shape the world. The challenges are complex, yet the creative, methodological, and ethical responses presented here offer compelling paths forward, grounded in empathy, evidence, and imagination.

Pescara, Italy Stefania Camplone
Genova, Italy Claudia Porfirione

Organization

Conference Organizer

EIDD—Design for All Europe

Conference Partners

Organizing Partner: City Council of Vila Nova de Gaia
Institutional Partners: Fundación ONCE
Tourism Authority of Porto

Organizing Committee

Pepetto Di Bucchianico, Italy
Maria Da Graça Guedes, Portugal
Markus Haas, Austria
Pete Kercher, Italy
Nuno Sa Leal, Portugal

Committee of Honor

President of the Republic of Portugal, Professor Marcelo Rebelo de Sousa
President of the Chamber of Deputies AR, Pedro Aguiar Branco
Prime Minister of Portugal, Luis Montenegro
Ministry of Defence of Portugal, Nuno Melo
Mayor of Vila Nova de Gaia, Eduardo Vitor Rodrigues
Mayor of Odessa, Gennady Trukhanov
Ambassador of City of Odessa in Italy, Attilio Malliani
Mayor of Porto, Rui Moreira
Mayor of Matosinhos, Luísa Salgueiro
Director of Accessibility and Innovation (Fundación ONCE) and of Accessible EU, Jesús Hernández Galan

Scientific Committee

Zuzana Ceresnova, Slovak University of Technology in Bratislava, Slovakia
Pepetto Di Bucchianico, University of Chieti-Pescara, Italy
Beata Fabisiak, Poznan University of Life Sciences, Poland
Rama Gheerawo, Royal College of Art, UK
Maria Da Graça Guedes, Minho University, Portugal
Jasmien Herssens, Fourmind, Belgium
Jannicke Hølen, Doga Design and Architecture, Norway
Claudia Porfirione, University of Genova, Italy
Janice Rieger, Griffith University, Australia
Nuno Sa Leal, University of Trás-os-Montes e Alto Douro, Portugal
Marta Więckowska, Academy of Fine Arts in Katowice, Poland

Contents

Academic Research: Spaces and Environments

Cities for All—Urban Aspects of Universal Design

Karol Görner, Lucia Štefancová, and Bohumil Kováč

Abstract

The issue of universal design needs to be perceived comprehensively in all spatial dimensions. Buildings and constructions designed in accordance with the principles of universal design are always situated in a certain urban context in accordance with the broader concept and spatial planning documentation. From a legislative point of view, spatial planning documentation and its functional and spatial regulation have a fundamental impact on buildings and constructions in the territory. It is therefore important to define the basic principles and tasks of universal design for individual levels of spatial planning documentation, especially at the level of the municipality and zone, and thus create conditions for the use of the territory of cities and municipalities by all residents. In urban development concepts, it is mainly about promoting the thesis of short-distance cities and creating a system of public spaces that can be used by everyone. The research focuses on the comparison of knowledge about universal design in zonal, municipal and regional contexts. The goal of the research is to transfer this knowledge into a set of recommendations for practice and the emerging Slovak legislation in the field of universal design and spatial planning. The outputs of research include the proposal of principles of urban planning and planning at individual levels of spatial planning (region, municipality, zone). This research was prepared for the National Project: Support of Universal Design.

Keywords

Universal design · Urban design · Spatial planning

1 Introduction

Universal design, alternatively Inclusive design (UK) or Design for all (EU), is one of the fundamental pillars of design and its fundamental principles are also enshrined in strategic documents such as Agenda 2030 for sustainable development, Urban Agenda for the EU, HABITAT III—The New Urban Agenda or the Convention on the Rights of Persons with Disabilities and Optional Protocol.

"Universal design" means the design of products, environments, programs and services to be usable by all people, to the greatest extent possible, without the need for adaptation or specialized design. "Universal design" shall not exclude assistive devices for groups of persons

K. Görner (✉) · L. Štefancová · B. Kováč
Slovak University of Technology in Bratislava, Faculty of Architecture and Design, Námestie slobody 19, Bratislava, Slovakia
e-mail: karol.gorner@stuba.sk

S. Camplone and C. Porfirione (eds.), *Design for All and the Future of the Planet*, Design For Inclusion 6, https://doi.org/10.1007/978-3-032-21769-1_1

with disabilities where this is needed" (United Nations, 2008).

Mace et al (1991) articulated seven basic principles of universal design: equivalence of use; flexibility of use; simplicity and intuitiveness of use; perceptual information; minimization of risk; minimization of physical effort, universal dimensions of space.

By 2050, 6.5 billion people are expected to live in cities, of which approximately one billion (15%) will be people with disabilities (WHO/World Bank, 2011). In Slovakia, people with disabilities make up 30% of the population (ŠÚSR, 2015).

Universal design in cities primarily concerns buildings and public spaces, which are always situated in a broader urban context. For this reason, it is essential to consider the broader urban aspects of universal design.

"Urban design is a discipline at the intersection of artistic creation, technical design and scientific work related to the creation of human settlements, but especially to the creation of cities. Simplistically, urban design means the creation of city-forming structures or the spatial framework of architecture" (Štefancová & Görner, 2024).

The urban aspects of universal design can be identified in different urban dimensions. The basic dimensions are the level of the region, settlement, zone and the level of the specific urban-architectural details of the design of public spaces.

Public spaces have a specific position in urbanism. Gehl (2000) sees public spaces as an important indicator of the life of city residents and visitors.

Public spaces are "places owned or used by the public, accessible and welcoming to all, free of charge and without profit motive. Public spaces are a key element of individual and social benefit, places of community social life, an expression of the diversity of their common natural and cultural wealth and the basis of their identity" (Garau & Lancerin, 2015).

A key factor in universal design is the accessibility and use of public space, which Rollová & Čerešňová (2015) understand as a human-legal, social and economic principle or way of providing access to the physical environment, to transport, to information as well as to resources and services for all.

2 Methodology

The issue of universal design is examined in four basic spatial dimensions: region; settlement; zone; urban-architectural detail.

The subject of the research is to identify the key urban aspects influencing urban development from the perspective of universal design in each spatial dimension. These aspects are discussed in the individual chapters of the paper.

The aim is to define the principles of urban universal design and thus highlight the importance of pursuing this issue to fulfil the ambition of achieving a sustainable society.

3 Regional Dimension

The region is a closely interconnected and hierarchically organized area with a wide range of functions that provide for all the needs of its inhabitants. The region experiences significant mobility daily.

The regional aspect of spatial planning in Slovakia is covered by three basic instruments in the form of spatial planning documentation at supra-regional and regional level: the Concept of Spatial Development of Slovakia; Spatial Plans of Regions; Spatial plans of micro-regions.

At the regional level, from the perspective of universal design, we can identify barriers: spatial-temporal (mobility time and distance) and physical (constraints and barriers).

3.1 Distribution of Functions in the Territory

In terms of the distribution of functions of regional importance, the region ideally represents a hierarchical system represented by development axes and centers of population in which important functions are concentrated. In this way it is possible to guarantee predictability, conceptuality, equity and universality of the distribution of functions. The livability of the landscape is also an important issue in regional planning, closely linked to recreational opportunities for all.

3.2 Forms of Mobility in the Region

Public transport in the Region

The region is experiencing population shifts in search of functions of regional importance, which are concentrated mainly in the regional center and in the more important towns located in the centers of population and on development axes. Part of the functions of regional importance are located outside the centers of population and development axes. These are mainly sports and recreation areas and protected natural areas. From the point of view of universal design, it is desirable to ensure that functions of regional importance are accessible by the widest possible range of transport modes in the shortest possible time. At the regional level, the provision of public transport complemented by a system of cycling and walking routes is key. Visitors and users of regional facilities are often those who rely on public transport (students, pensioners, parents with children, people with disabilities, patients undergoing treatment or people with other illnesses or injuries).

From a universal design perspective, it is ideal if regional public transport and local public transport form one integrated system with a minimum number of transfers.

Cycling in the Region

The bicycle is a sustainable and accessible means of transport for a wide group of people. At the regional level, the development of cycle routes is of recreational importance, but also enables distances between settlements and their functions to be covered. Although cycling is not accessible to all, cycle routes can also serve people in powered wheelchairs. Their profile should be dimensioned accordingly.

It is desirable to ensure that cycling and walking routes in the countryside are provided with comfortable and non-salubrious surfaces, by planting accompanying greenery (shading, windbreaks), choosing routes with gentler longitudinal gradients and with sufficient resting places.

4 Settlement Dimension

The basic instrument of spatial planning at the settlement level is the municipal spatial plan. In the settlement dimension, we perceive public spaces as a continuous whole, in which individual links are in a hierarchical arrangement (squares, embankments, streets, boulevards, urban parks, lower-order public urban spaces, inner neighborhoods). Connectivity and hierarchy in the organization of public spaces are key factors in universal design at the settlement level.

4.1 The Settlement of Short Distances

A settlement based on the principle of a compact city has a high potential to provide accessibility to amenities and settlement green space for all groups of residents.

In a short-distance settlement system, closely interlinked attributes must be met:

- high intensity of land use (increasing the efficiency of public transport and promoting a functionally mixed area)
- hierarchical distribution of amenities within the city (concentration along a network of focal public spaces contributes to the formation of local centers and increasing the efficiency of the urban structure)
- uniform hierarchical distribution of recreational green space within the city (availability of green space with emphasis on its quality)
- priority for walking, cycling and public transport

4.2 Internal Barriers to Settlements

In terms of movement restrictions, natural and artificial barriers can be identified in settlements. A key aspect of the City for All concept is the overcoming of artificial and natural barriers within settlements.

Natural barriers to settlements result from landscape features, in particular: impassable natural features (watercourse, etc.), morphological conditions (slopes and orientation to the cardinal points), weather influences (e.g. direction of prevailing winds). Natural barriers can be eliminated largely by appropriate technical solutions, by ensuring the highest possible number of points at which the barrier can be crossed to increase connectivity (see Fig. 1).

Artificial barriers in the settlement, on the other hand, are mainly caused by non-conceptual construction or errors in spatial planning: linear constructions (surface railway lines, road communications), impassable urban structure (large urban block, large, fenced area). In the case of artificial elements, the barrier effect should be avoided in the first place, e.g.: In the case of industrial zones and large enclosed areas, their location within the city should be carefully considered in advance, so that they minimally disrupt the natural operational links of the settlement structure.

5 Zonal Dimension

The zonal dimension of urban design and spatial planning is primarily concerned with the regulation of construction in the territory, which also includes the construction of public spaces. The zoning plan, as an instrument of spatial development and at the same time a basis for building permits, thus has a direct impact on the formation of accessible public spaces in terms of their universal use through regulation.

5.1 Short Distance Zone

The concept of a short distance town at the level of the residential area is primarily based on good accessibility to key points of interest (doctor, school, nursery, post office, grocery, local center, public transport stops, park). The distance of key points of interest from the residence should not exceed 500 meters (250 meters in the case of a nursery), which is equivalent to approximately five minutes walking distance.

To ensure the efficient operation of public transport, the viability of local amenities and pedestrian accessibility for as many residents as possible, an effective minimum population density of 150 inhabitants per hectare is needed.

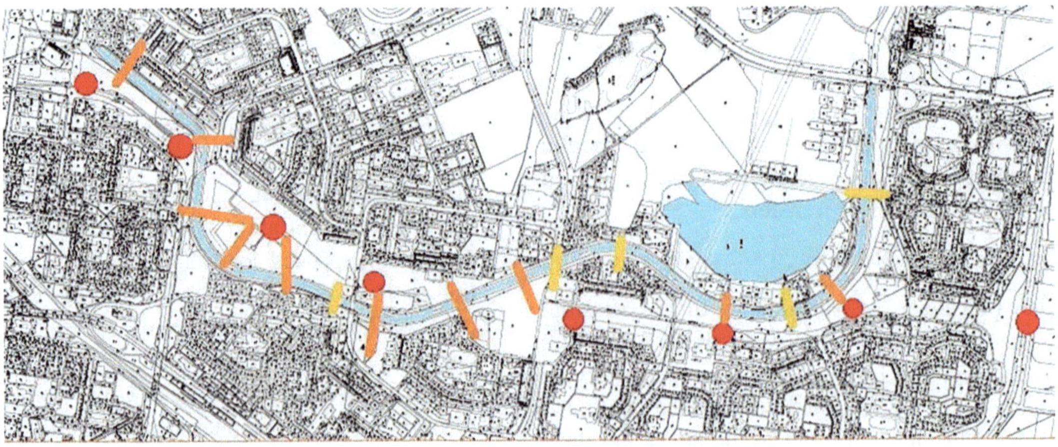

Fig. 1 Proposal to supplement the footbridge system to increase the connectivity of the area. *Authors* Bohumil Kováč & Karol Görner & Lucia Štefancová

At the same time, it is necessary to stick to a rational form of development and encourage pedestrian movement by shortcuts (passages), shading elements, elements improving the microclimate of the environment and overall well-being.

5.2 Compositional Aspects of Universal Urban Design of Zones

For the sake of aesthetic effect and orientation in space, it is important to pay attention to the compositional principles of material-spatial solutions when designing the urban study and the zoning plan of the zone. Kevin Lynch's (2004) research shows that the most important aspects of the environment in terms of orientation in space are:

- Roads (legible linear corridors of people's movement through space);
- edges (linear features forming a boundary in space—a barrier to pedestrian movement);
- areas (parts of the territory with a typical character);
- nodes (places of concentration and crossing where it is possible to enter);
- significant features (intensely perceived features).

In addition to the 5 phenomena of orientation described above, the urban composition uses a system of vistas, viewpoints and vantage points to further support wayfinding as one moves through the structure.

A properly composed urban structure helps to move intuitively through the area without the need for detailed knowledge of it, which is appreciated not only by visitors but also by people with reduced orientation skills.

5.3 Permeability of the Urban-Architectural Structure

The concept of universal design takes shape more easily in settlements on flat land than in settlements where public spaces are in a sloping position. Within a settlement with exposed terrain, vertical elevation can be overcome by universal design in various ways: ramps, ramp walkways and modified slopes; exterior lifts, escalators, elevators or travelators; suspension railways or funiculars. Overcoming terrain by stairs is of a barrier nature, which can be partially eliminated by moving platforms on handrails.

Barriers that create an impassable urban fabric or landscape significantly lengthen walking routes. The new urban structure should therefore be designed based on a rigorous spatiotemporal analysis of anticipated future pedestrian movement so that routes are as short as possible. It is desirable to guide permeability within the binding regulation of the zoning plan. The compact development of the urban block should not create a barrier to pedestrian movement of more than 200 meters.

The permeability through the urban structure depends primarily on the concept of the layout of public spaces (streets, squares,...) and additionally on the layout of complementary urban elements of pedestrian mobility (passages and passages through the urban structure). The importance of seeing the interior and exterior of the urban fabric, as a set of public urban spaces, courtyards and the ground floor of buildings that are open to the public, was already pointed out by Nolli (1748) in his black and white maps of Rome.

The importance of shortening walking distances lies in saving time, conserving physical energy, and eliminating the need to use individual car transport, which is particularly appreciated by the elderly, children and people with reduced mobility and orientation.

5.4 Designing Walking Routes for People with Disabilities

To create a city of short distances in the spirit of universal design, it is essential to think about the concept of a network of barrier-free routes linking important parts of the settlement formations. In addition to the broader concept of barrier-free routes, the zoning plan also provides scope for specifying the guiding lines for the visually impaired.

Guidelines in the public space should form a coherent system, which can be regulated in the zoning plan according to the type of guideline.

A natural guideline consists of the line of contact of a building wall with the plane of the sidewalk, the curb of a pedestrian street at the interface with a lawn, or the interface of two distinctly different tactile lines.

An artificial guideline is drawn loosely in a straight line through the space and consists of a band of embossed and color-contrasting surface.

From an urban design point of view, care should be taken to minimize the length of the route, avoiding areas where water collects (valleys) and consistently highlighting the change of direction of the route of the guideline.

From the point of view of people with reduced mobility, the design of public spaces should be particularly concerned with overcoming differences in height. A universal public space is one in which each stair step is accompanied by a ramp.

5.5 Accessibility, Safety and Comfort of Pedestrian Crossings

The importance of the design of pedestrian crossings over the roadway lies in ensuring maximum pedestrian safety. For a pedestrian crossing to be functional, it must first and foremost meet the accessibility condition. In addition to the standard measures to de-clutter crossings (navigation for the visually impaired, making the crossing accessible for the mobility impaired), it is advisable to minimize the length of pedestrian crossings. This can be achieved by:

- by reducing the number of lanes at the crossing point, which also means slowing down traffic but crossing the crossing more quickly;
- by narrowing the roadway, which also means slowing traffic but crossing the crossing more quickly;
- in the case of long crossings, the insertion of a divider island of sufficient width, optimally accompanied by a narrowing of the lanes at the crossing point.

6 Special Public Space Issues

At the scale of public space, a whole range of steps can be taken to shape a more inclusive space for all. One of the fundamental attributes of public space should be its safety. In particular, the impact of the weather and the technical design of key details on pedestrian routes should be kept in mind when it comes to safety.

In a period of global warming and related climatic extremes, it is important from the point of view of the more weather-sensitive people (seniors, children) to pay attention to the design of public spaces working with greenery, water and shading elements that help to create pleasant conditions for staying in the public space.

The application of art in public space can also help to make the stay more pleasant. An excellent example of an inclusive artistic element are models of buildings or urban structures. Their purpose is also to bring the visually impaired closer to the basic architectural forms of a building or the urban layout of a space by touching information in Braille (see Fig. 2).

A specific problem that needs to be tackled in public space for the sake of inclusion are mental barriers based on personal experience, subconscious fears and anxieties. These include, for

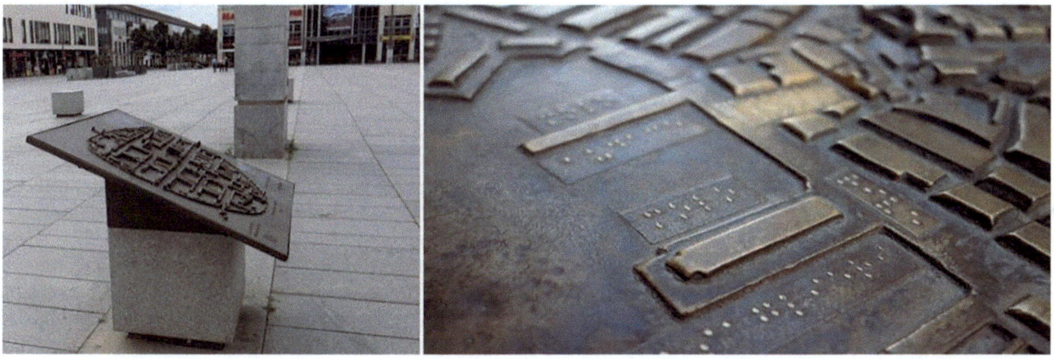

Fig. 2 Example of placing models of urban structure in a public space, left: Krems—the tilt of the model allows for a more comfortable position when touching and is better perceptible even from a wheelchair, right: Trieste—the model has directly integrated names of significant objects and streets in Braille. *Authors of the photo* Bohumil Kováč & Lucia Štefancová

example, darkness, confined spaces, busy traffic and so on. The basic solution to these barriers is to design public spaces that are visually clear, sufficiently lit and participatively planned.

7 Conclusion

Universal design cannot be seen only through design solutions for specific user details. Opportunities to make the user environment more universal, inclusive and sustainable need to start looking at the broader urban context.

Presented article include the proposal of basic principles of urban planning and planning at individual levels of spatial planning (see Table 1).

Following these principles in the elaboration of spatial planning documentation and design of public spaces creates a fundamental prerequisite for the creation of an inclusive environment in the standard of universal design.

Table 1 Basic principles of urban aspects of universal design in spatial dimensions

Spatial dimension	Basic principles of urban aspects of universal design
Region	Hierarchical distribution of functions; ensure accessibility to amenity centers and recreational sites; well-developed public transport and cycling routes
Settlement	Well connected, functionally mixed and compact city structure; good accessibility of amenities and green infrastructure; support of walking, cycling and public transport; elimination of natural and prevention of artificial barriers
Zone	Accessibility of basic amenities in walkable distance; ensure sufficient population density; well-connected hierarchical system of public spaces; support of orientation via urban composition principles; system of walking routes for disabled people
Urban-architectural detail	Safety details (elimination of potential risks); creating pleasant conditions for staying in public space; creating elements supporting inclusions; suppression of mental barriers

References

Garau, P., Lancerin, L., & Sepe, M. (2015). *The charter of public space.* LISt Lab.

Gehl, J. (2000). *Život mezi budovami.* Nadace Partnerství.

Lynch, K. (2004). *Obraz mesta.* Bova Polygon.

Mace, R. L., Hardie, G. J., & Place, J. P. (1991). Accessible environments. Toward universal design. In W. F. E. Preiser et al. (Eds.), *Design Interventon toward a more humane architecture.* Van Nostrand Reinhold.

Nolli, G.B. (1748). Nuova pianta di Roma. Map, Rome.

Rollová, L., & Čerešňová, Z. (2015). *Univerzálne navrhovanie objektov komunitných sociálnych služieb.* Implementačná agentúra MPSVR SR.

Štefancová, L., & Görner, K. (2024). *365 pojmov z urbanizmu.* SPEKTRUM STU.

ŠÚSR (2015): Vybrané indikátory sociálnej situácie osôb so zdravotným postihnutím. https://www.employment.gov.sk/files/slovensky/rodina-socialna-pomoc/tazke-zdravotne-postihnutie/vybrane_indikatory_socialnej_situacie_osob_so_zp.pdf, Reterived 2024/11/22.

UNITED NATIONS (2008): Convention on the rights of persons with disabilities and optional protocol., https://www.un.org/disabilities/documents/convention/convoptprot-e.pdf, Reterived 2024/11/22.

WHO/World Bank (2011): The Inclusion imperative: Towards disability-inclusive and accessible urban development. www.cbm.org/Publications-252011.php, Reterived 2024/11/22.

The Role of Research Methods in Describing User Experiences to Inform Future Design

Emil Ballegaard, Roberta Cassi, and Masashi Kajita

Abstract

Obtaining relevant information about usage is crucial for designing environments that accommodate diverse user preferences and needs. Despite well-documented research methods for data collection and analysis, their role in guiding design is often overlooked. This study investigates the role of different research methods in describing and processing user experiences to inform future designs. Researchers at the Cluster for Spatial Inclusion at The Royal Danish Academy collaborated to share examples of research methods used to collect and process descriptions of users' interactions with built environments. Reflections on these methods and outcomes form the basis of a comparative study highlighting methodological characteristics related to data collection, data analysis, and knowledge application. The study shows that the role of methods in gathering descriptive information about user experience largely depends on the conceptualisation of users and their role in the production and application of knowledge. By breaking down the processes into smaller steps, the study highlights the similarities and differences in the methods applied. Besides methodological variations and their imbrications, the paper presents a collective discussion of gaps and biases identified in the Cluster's workflows. The paper concludes by stating that increased awareness of the implications of methodological choices in research design development could potentially enhance the validity of knowledge applied to accommodate the preferences, needs, and desires of users.

Keywords

Design studies · Research methods · Spatial inclusion · Usage · User experience

1 Introduction

Designing environments that cater to a diversity of users requires an understanding of how they currently use, and would like to use, these environments (Imrie & Hall, 2003). In other words, understanding the user-environment interaction is fundamental to creating environments that accommodate a broader spectrum of users (Boys, 2014). Acquiring pertinent information about the use of the environment is, therefore, imperative for developing designs that effectively support and enhance user

E. Ballegaard · R. Cassi (✉) · M. Kajita
Royal Danish Academy, Copenhagen, Denmark
e-mail: rcas@kglakademi.dk

E. Ballegaard
e-mail: emba@kglakademi.dk

M. Kajita
e-mail: masashi.kajita@kglakademi.dk

S. Camplone and C. Porfirione (eds.), *Design for All and the Future of the Planet*, Design For Inclusion 6, https://doi.org/10.1007/978-3-032-21769-1_2

experiences (Hendren, 2020) (Cupers, 2013). In architecture and design, studies of usage are critical, as they provide the necessary information for architects, designers, and other decision-makers to accommodate users' preferences, needs, and desires. We can recognise, for this purpose, that research methods serve three primary functions: first, to collect information about users and their interactions with built environments; second, to process and analyse this information to develop knowledge; and third, to transfer this knowledge to inform how design processes can respond to future interactions.

People-centred design concepts, such as Universal Design, Design-for-All, and Inclusive Design, all respond to the widespread exclusion of users from built environments (Persson et al., 2015). The concepts share an emphasis on user's perspective regarding the use of designs as a key element in accommodating the users—regardless of bodily, cognitive, or social conditions. (Imrie & Hall, 2003; Hendren, 2020; Steinfeld & Maisel, 2012). To gather subjective and contextual information about usage, qualitative and participatory research methods are widely applied among people-centred design practitioners—although this is not always the case (Steinfeld & Maisel, 2012; Kajita, 2020). While these methods excel at collecting information that describes particular users, their experiences, and specific use events, the generalisation of this information and its effective transfer to design processes are questioned.

There are several challenges for architects and designers in implementing such information. One challenge is the difficulty (if not impossibility) of transferring knowledge from past contexts and situations into future applications due to their inherent contextual nature (Merit et al., 2022). Additionally, this issue complicates trial-and-error methods, which are often employed by architects and designers, since many aspects of usability can only be evaluated over time (Lawson, 2006). Another challenge exists in the incompatibility between information that describe subjective and contextual features and the information which can be instructive to guide decision-making during the design process (Lawson, 2013). This problem is linked to the fundamental differences between knowing in traditional research, which aims to describe, explain, or predict phenomena (Lawson, 2002) and *designerly ways of knowing* (Cross, 1982), which focus on proposing future forms and arrangements.

Although significant efforts have been made to understand and bridge these *gaps* (Van de Wejer et al., 2014; Van der Linden et al., 2019), further research is needed to explore how subjective and contextual information can guide designers toward solutions that effectively address users' preferences, needs, and desires. At the Royal Danish Academy, the Cluster for Spatial Inclusion (hereafter referred to as "Cluster") is committed to explore ways to bring lived experience of persons into a vocabulary of space makings for architects and designers.[1] This paper presents a collaborative effort by researchers at the Cluster to share examples of research methods used in various projects in order to: first, collect information about users; second, build knowledge from this information; and third, communicate the knowledge. By comparing these examples, the aim is to demonstrate how researchers at the Cluster have approached the aforementioned knowledge gap and explore potential ways forward.

2 Background

To provide the reader a comprehensive overview for this study, the following section presents short descriptions of the selected Cluster's research projects, outlining research aims, methods applied to gain information about users (or those who use the information of users), and results.

P1—Places and Small Pleasures explores how environments can foster moments of pleasure for disabled people by exploring the relationship between environmental and personal

[1] https://royaldanishacademy.com/en/cluster-spatial-inclusion/about-cluster-spatial-inclusion.

dimensions. By interviewing four disabled persons, this study collects their personal accounts and reflections on how certain environments contributed to their feeling of pleasure. Through the use of narratives, experiences of the four participants are represented in the form of drawings and film to articulate their everyday pleasures and the role that the built environment plays in fostering these moments.

P2—Disability, Experience and Architecture investigates how the built environment can support the participation of individuals with impairments in sports and leisure activities. By interviewing architects and users of two sports and leisure buildings, the study identifies how characteristics of the built environment enable users while performing cognitive, physical or social activities. The findings offer performance-based strategies to increase architects' awareness on the influences of design on users' performances.

P3—Uncertainty in Digital Design explores how data describing both experiential and technical aspects can be applied in digital architectural design processes. Using a conceptual data model, data from multiple sources is integrated in a graph holding knowledge related to both build environments and to users' perceptions of those environments. The data is applied to inform rule-based generative design processes. Preliminary results indicate that a thorough data strategy and careful distribution of tasks between human designers and computers can enable workflows that engage with multiple future use scenarios and are sensitive to user statements.

P4—Coordinating Inclusive Future Public Spaces in Architecture examines how disabled users are considered in the design processes of public spaces. Using participatory ethnography, the project studies the design process and representational design *devices* employed during the development of future urban scenarios. Findings from this study offer recommendations for the use of representations of marginalised users in design processes.

To further illustrate the methodological approaches of each project, the following table summarises the research methods employed for data collection, data process/analysis, and knowledge dissemination.

3 Methodology

The selected Cluster projects served as cases for a comparative study (Jordan & Gross, 2011; Bartlett & Vavrus, 2016). The study aimed to elucidate the projects' methodological characteristics related to the collection and analysis of data describing various aspects of usage, as well as the application of knowledge built from this data. In doing so, we questioned how the application of knowledge to accommodate users through design is reflected in the choices of collecting and processing of data from users? During internal meetings in the Cluster prior to the comparative study, projects were presented and discussed to understand the role of the research methods employed. This led to the recognition of a shared chronology of research "steps" among the projects, as shown in Table 1. By using Table 1 as structure, the comparative study seeks to bring forth the projects' differences and similarities relating to both individual and multiple research steps.

4 Results of Comparative Study

4.1 Data Collection

All research projects examined in this study employed interviews to gain information about users. However, interviewees selection, interview procedures, chosen locations for conducting interviews, and the methods used to capture and preserve the collected data vary significantly from project to project, relating to their research designs and questions. In three of the four projects (P1, P2, P3), focus was on the understanding of users' experiences and interactions with the spatial settings and characteristics. For this reason, interviewees selected were individuals considered as primal users of the built environment. In P1 and P2 participants were deliberately chosen to represent a diverse

Table 1 Methods employed by the selected cluster's projects for data collection, data analysis and knowledge application (*Source* Authors)

<table>
<tr><th></th><th>P1
Places and Small Pleasures</th><th>P2
Disability, Experience and Architecture</th><th colspan="2">P3
Uncertainty in Digital Design</th><th>P4
Coordinating Inclusive Future Public Spaces</th></tr>
<tr><td colspan="6">Data collection</td></tr>
<tr><td>Intervieew</td><td>Users with diverse bodily conditions, and of different gender and age.</td><td>Users with mobility, visual or hearing impairments</td><td colspan="2">Elderly users of domestic architecture.</td><td>Landscape architects</td></tr>
<tr><td>Process</td><td>In depth semi-structured interviews</td><td>Iterative in depth semi-structured interviews</td><td colspan="2">Iterative in depth semi- structured interviews</td><td>Participatory ethnographic fieldwork with unstructured interviews</td></tr>
<tr><td>Location</td><td>Setting provided by researcher</td><td>In-situ, go-along</td><td colspan="2">In-situ</td><td>Architects' office</td></tr>
<tr><td>Documentation</td><td>Audio-video recording</td><td>Audio-video recording</td><td colspan="2">Audio-recording</td><td>Field notes
Audio-recording</td></tr>
<tr><td colspan="6">Data analysis</td></tr>
<tr><td>Coding</td><td>Coding based on pleasure related themes</td><td>Coding based on person-environment analytical model</td><td colspan="2">Coding using a formal ontological model</td><td>Coding based on employed users' representations and design process temporality</td></tr>
<tr><td>Knowledge</td><td>Narratives describing individual experiences of pleasure in space</td><td>Enabling mechanisms offered by the environment to the user for performing activities</td><td colspan="2">Graph database communicating relations between built environments and users' experience of those environments</td><td>Roles of representations in the accommodation of future marginalised users' preferences and needs.</td></tr>
<tr><td colspan="6">Knowledge application</td></tr>
<tr><td>Research outcome</td><td colspan="2">Drawings and film representing the experience of pleasure</td><td>Performance-based design strategies</td><td>Data that can be used to represent usage in digital architectural design</td><td>Inclusivity guide with reflective questions for practitioners</td></tr>
</table>

range of physical conditions and abilities. In P3, without imposing any functional criteria, older people living independently were selected. Differently, in P4, the interviewees were professionals, in this case, landscape architects, who served to provide information about how interests and representations of disabled users are represented in an actual design process.

Semi-structured interviews were carried out in P1, P2, and P3 with sessions lasting one to two hours, sometimes spanning multiple sessions. P2 and P3 employed an iterative technique, involving a series of interviews with the same participant to delve deeper into user-environment interactions and provide rich descriptions of settings and the users' experiences. Here, information collected throughout the first interview, provided a baseline that informed following sessions with new questions for semi-structured interviews. In P4, on the other hand, participatory ethnographic fieldwork was conducted through a more immersive and continuous interview process, where practitioners were questioned while observing their workflows and ways of including various representations of disabled users in the design process.

Carried out in sport and leisure facilities and in private homes respectively, the interviews of P2 and P3 were conducted in-situ, providing context and reference for users to express details about their usage. For P1, instead, main requirement for the interview location was that the participants should feel comfortable to open up and talk freely. In P4, interviews with professionals were conducted in the architects' office. This allowed the researcher to ask questions while observing their workflows. Various media were employed across the projects for documenting the interviews, including audio recordings, audio-video recordings, and note-taking. Especially in P1, the video-recordings of the interviews provided audio-visual material eventually employed for the production of a short film. In P4, data registration during participatory fieldwork included around one hundred hours of architects' meetings audio-recording linked to fieldnotes and time references (Merit et al., 2022).

4.2 Data Analysis

In each of the research projects data collected was processes and/or analysed to address specific research objectives. A coding approach was employed by researchers in all projects to anchor information to relevant themes, categories, entities and develop the understanding of the object under study. In P1, grounded theory was applied to guide the selection of codes, and identified key themes and aspects related to the positive spatial experiences reported by the interviewees were organised by using diagrams, drawings, and storyboards. In P2, the interviews were systematically coded according to a developed *usability model* that utilises three categories: person, environment, and activity (Cassi et al., 2021). Similarly, in P3, a *formal ontological model* was created to structure the coding of heterogenous data according to classes, namely subclasses of *object* and *event*. To identify the presence of representations of users with disabilities in a design process, the coding in P4 tracks when (timestamp) disabled users are mentioned (and when they are notably absent) among the professionals.

The knowledge developed throughout the data analysis in these research projects covered various aspects of users' experiences of the built environment. In P1, experience of the built environment was examined through the four participants' narratives of their positive experiences, showcasing the unique relations of various attributes for creating the feeling of pleasure. Similarly, P2 highlights the potential enabling mechanisms offered by the designed characteristics of the built environment (Cassi, 2023) and revealed relevant influences of design to users' possibility to perform in space. In P3, the structured data highlighted opportunities and challenges in the model's structure for describing both specific and general objects and events, as interpreted by users and professionals. Finally, in P4, a timeline was developed to register how and when disabled users are explicitly or implicitly represented in the design process. This mapping indicated roles of representational *devices,* for each design stage and in reference to different levels of application, in contributing to the inclusion of disabled users in the design process (Merit et al., 2022).

4.3 Knowledge Application

In P1, a film was produced which shows a scenario of pleasure re-enacted by each of the four participants, with their vocal descriptions serving as voiceovers (Kajita et al., 2024). The film not only provided a more engaging and accessible way to present research findings but also facilitated audience discussions during public exhibitions and screenings. P2 developed performance-oriented design strategies to guide architects on how environmental characteristics—such as materiality, dimensions, organisation, lighting, and acoustics—can support activities of individuals with mobility, visual, and hearing impairments (Cassi, 2023), through which architects could develop solutions that better support and stimulate users' activities. P3 sought to utilise integrated information from users and professionals to guide the design of architecture that considers multiple future use events. Visually presented multiple use scenarios involving users' narratives and the objects of usage challenges assumptions among professionals and support dialogue between users and professionals (Ballegaard et al., 2024). Similarly, in P4, the analysis of the devices used for representing disabled users in a design process led to the creation of an inclusivity guide. The guide comprises a series of questions directed to architects and other decision-makers at different work stages in design processes to foster awareness of potential exclusion of disabled users.

5 Discussion

Transferring and transforming information from interviews into understandable knowledge that can be interpreted and applied in contexts of design presents several challenges (Persson et al., 2015; Merit et al., 2022; Jordan & Gross, 2011). The projects compared have all sought to address these challenges, aiming to deliver valid knowledge, about and from users, enabling receivers to act—whether by increasing their awareness, designing inclusive spaces, or engaging in other activities. This shared aim reflects writings on the Cluster's website in which we state that "we explore ways to bring lived experience of persons into a vocabulary of space makings for architects and designers." However, scrutinising the results of the comparative study has led us to question both the validity of this statement and the durability of the presented research designs in achieving this endeavour: Is it even possible to transmute lived experience into a vocabulary without losing essential properties of the experiences? This is regarded an open question, which the compared projects have dealt with by proposing various "vocabularies" of knowledge that can be utilised to mitigate exclusion of users.

An aim shared by all projects appears to be enabling receivers of knowledge to gain "awareness" of aspects of users and usage through their interpretation of each project's respective knowledge. However, the knowledge presented in each project suggests significant differences in the receivers' processes of gaining awareness through interpreting the knowledge. Based on the interview data, P2 and P4 presented generalised knowledge in the form of strategies and process guides that professionals can utilise during design processes in different contexts to gain awareness of potential user exclusion. By contrast, and for similar reasons, P3 segmented and connected data from users and professionals, creating a database structure that enables designing professionals to engage directly with the segmented interview data. In a different approach, P1 communicated knowledge through film, aiming at a broader audience of both professionals and laypersons to increase general awareness of marginalised perspectives on the interplay between users and environments.

Although generalisations are inherently part of all the projects, the varying forms of generalisation spurred discussions among the Cluster researchers about how these might affect knowledge receivers and their (potential) designs. While no definitive conclusions emerged from these discussions, essential details related to the coding or classification of data across projects became evident. For instance, the linkage between the data codes and the final communication of knowledge, often shaped by decisions made during the arrangement of data collection, is largely overlooked. An example of this is the process guides presented in P4, which aim to prompt professionals at specific points during design processes, reflecting the timestamp-based coding of data collected in a working environment. Similarly, the knowledge graph presented in P3 can be queried by professionals using the coded objects and events mentioned by interviewees as points of reference. In this knowledge graph, the codes also respond to the in-situ interviews, in which interviewees reflected on the objects at hand and their roles in various events. In this sense, and perhaps obviously in hindsight, we recognised that the coding of data frames the knowledge available for the knowledge receivers to act upon.

This finding fundamentally calls into question the research design of all the projects, as users were absent from the processes of coding data, despite a shared emphasis on establishing a valid connection between what is relevant to users and the knowledge communicated. Furthermore, this finding also raises detailed and critical questions about the role of professional researchers and designers as gatekeepers of knowledge regarding users' preferences, needs, and desires. These questions suggest that the Cluster researchers implicitly define awareness as a one-directional state, where professionals are conscious of users and usage aspects. Instead, we believe awareness should be viewed as two-directional, incorporating users' consciousness of relevant aspects about professionals' work. This would include the aforementioned coding of data, the construction of user representations as particular or general, the differentiation of objects and events, and many other critical research and design decisions. Implementing awareness as mutual understandings between laypersons and professionals of what is relevant to each other, as an explicit goal in research and design processes, could lead to defining more active roles of users. This motivation reflects an established viewpoint among scholars in and around the field of architecture (Cupers, 2013; Hill, 1998; Oudshoorn & Pinch, 2003), including scholars in the field of people-centred design advocating for critical and intersectional approaches to the accommodation of users in architectural design (Boys, 2014; Hamraie, 2017). We believe that such approaches, in which usage is regarded as processes involving physical, bodily, psychological, and cultural aspects, could provide a viable framework for integrating the findings of this study into future research and design activities within the cluster. Hopefully, these activities will contribute to the advancement of valid and instructive knowledge about subjective and contextual aspects, thereby supporting architects and designers in accommodating users' preferences, needs, and desires.

6 Conclusion

Building knowledge to guide architects and designers in accommodating future users presents numerous challenges. Seeking to contribute to the engagement with active users in architecture and design, this paper presented a comparative study of four projects carried out within the Cluster for Spatial Inclusion at the Royal Danish Academy. Each project was examined in terms of: 1) how information about users was collected; 2) how knowledge was constructed from this information; and 3) how the resulting knowledge about users was communicated. The findings formed the basis for discussions within the Cluster. These discussions not only highlighted differences across the projects' approaches but also fostered a shared recognition of gaps in our current research on and with users. This reflection prompted us to critically question the approaches, concepts, and statements currently employed in the Cluster and to propose directions for future research aimed at implementing the study's findings.

References

Ballegaard, E., Kajita, M., & Foged, I. W. (2024). Narratives in generative architectural design: Exploring sequential functions and usage. In universal design 2024: Shaping a sustainable. *Equitable and Resilient Future for All.*

Bartlett, L., & Vavrus, F. (2016). *Rethinking case study research: A comparative approach.* Routledge.

Boys, J. (2014). *Doing disability differently: An alternative handbook on architecture, dis/ability and designing for everyday life.* Routledge.

Cassi, R. (2023). *Disability, Experience and Architecture: Towards Inclusive Sports and Leisure buildings.* PhD dissertation

Cassi, R., Kajita, M., & Popovic Larsen, O. (2021). User-environment interaction: The usability model for universal design assessment. In *In universal design 2021: From special to mainstream solutions.*

Cross, N. (1982). *Designerly ways of knowing. Design studies, 3*(4), 221–227.

Cupers, K. (Ed.). (2013). *Use matters: An alternative history of architecture.* Routledge.

Hamraie, A. (2017). *Building access: Universal design and the politics of disability*. University of Minnesota Press.

Hendren, S.: What can a body do?, (2020).

Hill, J. (Ed.). (1998). *Occupying architecture: Between the architect and the user*. Routledge.

Imrie, R., & Hall, P. (2003). *Inclusive design: Designing and developing accessible environments*. Taylor & Francis.

Jordan, E., Gross, M.E., Javernick-Will, A.N.: Use and misuse of qualitative comparative analysis. Construction Management and Economics, pp. 1159–1173 (2011).

Kajita, M. (2020). What kind of bodies do architects imagine when designing. In D. Hauderowicz & K. Ly Serena (Eds.), *Age-inclusive public space, Hatje Cantz Verlag* (pp. 116–119).

Kajita, M., Cassi, R., & Græsbøll Holm, P. (2024). Filmmaking as a tool for investigating the positive spatial experiences. In universal design 2024: Shaping a sustainable. *Equitable and Resilient Future for All*.

Lawson, B. (2002). Design as research. Arq. *Architectural Research Quarterly, 6*(2), 109–114.

Lawson, B. (2006). *How designers think*. Routledge.

Lawson, B. (2013). Design and the evidence. *Procedia—Social and Behavioral Sciences, 105*.

Merit, M. T., Kajita, M., & Krarup, J. M. (2022). Elusive users: The presence of physically disabled users within architectural design processes. *Architecture, 3*(1), 56–70.

Oudshoorn, N., & Pinch, T. (Eds.). (2003). *How users matter: The co-constructions of users and technology*. The MIT Press.

Persson, H., Åhman, H., Yngling, A. A., & Gulliksen, J. (2015). Universal design, inclusive design, accessible design, design for all: Different concepts—One goal? On the concept of accessibility—Historical, methodological and philosophical aspects. *Universal Access in the Information Society, 14*, 505–526.

Steinfeld, E., & Maisel, J. (2012). *Universal design: Creating inclusive environments*. Wiley.

Van de Wejer, M., Van Cleempoel, K., & Heynen, H. (2014). Positioning research and design in academia and practice: A contribution to a continuing debate. *Design Issues, 30*(2), 17–29.

Van der Linden, V., Dong, H., & Heylighen, A. (2019). Tracing architects fragile knowing about users in the socio-material environment of design practice. *Design Studies, 63*, 65–91.

VR Technology in Wayfinding Research—A Comparison of User Behavior in VR and RL Environment

Marta Więckowska and Patrycja Rudnicka

Abstract

Wayfinding is critical for designing user-friendly public spaces. However, the diversity of architectural solutions, building characteristics, and cultural contexts makes it difficult to establish universal guidelines for wayfinding design. Each case requires tailored research, and traditional methods like shadowing and observation can sometimes be challenging to implement. Additionally, testing new solutions demands creating physical prototypes, which is time-consuming and costly. This paper explores the use of virtual reality (VR) as a tool for wayfinding research. The study involved 19 participants searching for specific locations in a building with diverse architecture and lighting conditions in a both real-life (RL) setting and its VR replication created using 3D spherical images. The findings revealed that navigation behavior was similar across RL and VR settings. Although participants used various strategies, their choices were influenced by their personal preferences rather than the type of environment (RL or VR). The results demonstrate the potential of VR as a valuable tool for wayfinding research, providing insights into how users develop spatial orientation and navigate within complex environments. The VR environment, utilizing high-quality 360° 3D imaging, realistically and faithfully replicated the building characteristics. This approach is particularly well-suited for evaluating the logic and clarity of wayfinding systems. VR allows researchers to modify spatial views by adding, removing, or altering elements (e.g., signage or colors) without constructing physical prototypes. These capabilities make VR a promising tool for advancing wayfinding design and testing.

Keywords

Wayfinding behavior · Virtual reality · User experience · Route strategy

1 Introduction

Finding a way in a natural or built environment is essential to everyday human activity. To achieve this, we use spatial and environmental information to navigate to a destination (Lynch, 2005). Wayfinding, which can be defined as the experience of locating, following, or discovering a route through and to a given space, is both a cognitive and physical process (Symonds et al., 2017). Reading landmarks

M. Więckowska (✉)
Faculty of Design, Academy of Fine Arts and Design in Katowice, Katowice, Poland
e-mail: marta.wieckowska@aspkat.edu.pl

P. Rudnicka
Department of Theory and History of Art, Academy of Fine Arts and Design in Katowice, Katowice, Poland
e-mail: patrycja.rudnicka@aspkat.edu.pl

S. Camplone and C. Porfirione (eds.), *Design for All and the Future of the Planet*, Design For Inclusion 6, https://doi.org/10.1007/978-3-032-21769-1_3

and choosing the path is inherently complex; it demands locomotion and navigation, activating cognitive, sensory, memory, and decision-making systems along with motor functions (Montello & Sas, 2006). There are individual differences in wayfinding abilities determined by age, sex, culture, and psychological state (Jamshidi et al., 2020). Recently, researchers have emphasized that wayfinding is also a kind of social activity in which we are influenced by other people's actions or even by their mere presence and artifacts (Dalton et al., 2019).

Wayfinding is the process of finding your way to a destination in a familiar or unfamiliar setting using environmental cues (Farr et al., 2012). Planning and creating such cues and tools to help users easily navigate the space is at the heart of signage and wayfinding system design. Usually, a wayfinding system includes maps and plans, various forms of signage, and information kiosks/points. However, it might also include mobile apps, Bluetooth beacons or even staff services provided by guides. An adequately designed wayfinding system allows people not only to find a way faster and with fewer errors but also influences their confidence and perception of safety.

Designing a good wayfinding system is a multidisciplinary effort as it includes the analysis of possible user paths, understanding of the institution for which the system is designed including legal regulations, and making appropriate decisions about colors, forms, and materials choices. The wayfinding system is site–/building-specific; each system should be in tune with the building's character and function, cultural context, or diversity requirements while maintaining universal rules regarding legibility, signage placements in decision points, or legal requirements. Poor wayfinding design might have grave consequences, especially in busy spots like railway stations, airports, or hospitals. However, in calmer environments, it creates the image of the institution and might make the visitors' experience a delightful one or lead to frustration and loss.

Research during the prototyping and testing stages is crucial for successfully designing a wayfinding system. It is necessary to check assumptions about user paths, check the placement and content of signage, and test the general legibility and understandability of the system as a whole. It is not always easy as sometimes the wayfinding system is designed before the building is even erected. Designers use paper prototyping, but in such cases, usually only a limited number of versions can be tested. Such research is also time-and resource-consuming: testing new solutions requires producing and placing prototypes in the building and then preparing shadowing and observation experiments, which are not always feasible in RL environments. Virtual reality technology offers new solutions for prototyping and design testing (Al-Sharaa et al, 2022), however, it still requires learning more about the specifics of wayfinding in real-life and virtual reality (VR) settings.

1.1 Wayfinding Behavior in VR Environment

VR technology enables full immersion in a created reality. However, the technology is still not widespread due to costs and the need for additional equipment, such as VR glasses. Nevertheless, the potential of the technology is undeniable. It is often used in training that requires specific conditions, which are difficult to achieve in a laboratory setting (e.g., training for industry, medical procedures, and even developing interpersonal skills). Several studies have compared wayfinding behavior in real-world versus VR settings. In the past, when the technology was less developed, some differences in behavior in both conditions were noted (Haq et al., 2009; Skorupka, 2009); however, the cognitive processes and use of environmental cues were similar. In more recent studies, similarities in behavior, particularly in how individuals search for specific destinations across both conditions, were confirmed (Dong et al., 2022; Halik & Kent, 2021; van Beek et al., 2024). Goupil et al. (2023) suggested that the context of the information-seeking scenario may have influenced participants to exhibit similar responses in both environments. They also noted that poorly designed or confusing signage would hinder participants' ability to comprehend their surroundings, thereby slowing their learning processes in both scenarios. Ewart & Johnson (2021) and van Beek (2024) analyzed head movement in both conditions and revealed that in VR it is lower, but it didn't influence how participants performed searching for the destination.

A crucial element in those studies was the characteristics of the buildings. The aforementioned VR studies were conducted in more complex multi-level buildings, allowing for a broader generalization of VR results (Feng et al., 2022; Kuliga et al., 2020; van Beek et al., 2024), while previous research mainly focused on single or two-story buildings (Dong et al., 2022; Ewart & Johnson, 2021). Building features influence wayfinding performance and strategy choice in VR, especially when participants searched for destinations requiring movement between floors.

Finally, an important element when comparing the results for RL vs VR studies is the type of technology used in the experiments, as there is a variety of them ranging across various levels of immersion possible, from desktop solutions to head-mounted devices (HMD) or Cave Automatic Virtual Environment (CAVE). The immersion level is an important element influencing user behavior (Feng et al., 2022). Some technologies are also better suited to faithfully render reality. VR settings using 3D modeling programs usually had some visual differences in the interior design. In our research, we decided to use 360-degree photos, which allowed us to create a realistic replica of the space, including differences in lighting in corridors and staircases.

Nonetheless, as wayfinding is a complex cognitive and social experience, some research has focused on individual differences when using VR technology for research. Goodroe and colleagues (2025) compared the influence of age on wayfinding tasks, finding that younger participants outperformed older ones, but only in VR settings. Other studies indicated that there are mostly no gender differences, often related to gaming experience; however, in experimental VR conditions females were faster than males when finding a location (Vilar et al., 2014). Thus, the hypothesis that gaming experience might influence VR performance is not universal, especially when no special equipment such as joysticks is necessary. Finally, the role of co-presence of others has been explored recently. Co-presence, often an overlooked aspect while designing VR experimental environments, affected how participants decided what to do and how they viewed physical navigation aids such as signs and lighting (Yassin et al., 2021). It shows the importance of real-life rendering not only in terms of structural (levels, general faithfulness) and qualitative (lighting, details) design but also social elements (co-presence) in constructing experimental virtual environments.

1.2 Research Objectives

We had two objectives in this study. The first was to identify the similarities and differences in human wayfinding behavior between real-world and virtual reality (VR) settings. The second objective was to evaluate the space and identify factors that help or hinder finding the specific location in real-life and virtual environments.

The study's results will provide insights into human wayfinding behavior and the interaction between spatial characteristics and behavior. Understanding potential differences in behavior between VR and real-life settings will also aid in developing and planning future wayfinding research. Traditional research methods are often time-consuming, resource-intensive, and difficult to execute. VR-based research offers numerous advantages, but the extent to which human behavior aligns in both environments requires careful examination.

2 Method

We set up three experimental settings; the first group was tested in real-life conditions (RL-only), the second exclusively in the virtual setting (VR-only), and the third group was tested in the real world and then in virtual reality after at least six months.

Participants searched for three different rooms located on the first and second floors. During the real-life walks, they recorded their route using a 180-degree camera; in VR, the 180-degree recording was also used. The think-aloud protocol was used in both conditions, and shadowing was used in the real-life setting. At the end of this part of the experiment, participants were asked to fill in a questionnaire evaluating the space impression and the wayfinding experience.

2.1 Building and Destinations Selection

The criteria for choosing the building included complexity (multi-storey building, equipped with elevators and stairs), diversity (halls and corridors of different widths, various types of lighting), and the presence of at least a basic wayfinding system. The experiment took place in the building of the University of Silesia's Faculty of Law and Administration in Katowice.

The experiment's destinations were chosen to observe different wayfinding scenarios (locations on different floors—two on the second floor and one on the first floor, target indication—by number or function/office name) and users' behaviors (tendency to choose or not to choose specific strategies, i.e. stairs, elevators, or specific corridors, ability to learn the way and the way back).

2.2 Virtual Setting Preparation

The virtual walk was constructed using separate panorama photos linked logically, allowing users to move seamlessly from one place to another within the distance of a few meters.

The photos were taken using an 8 K camera to achieve the highest resolution and most accurate representation of reality in capturing the interior of the building to create the most immersive effect. The places where the photos were taken were chosen carefully, and the virtual walk covered all the places where the wayfinding information was placed. It also allowed for smooth switching of the panoramas. The test began with instructions on navigating the virtual walk and switching to the desired 360-degree panorama. To move to a different location, users needed to focus on a specific area of the image—indicated by a circle. Upon doing so, they would receive visual and auditory confirmation that the panorama was successfully switched.

2.3 Experimental Procedure

The experimental procedure included four parts: (1) an introduction at the front of the building: participants were informed about the experiment's aim and procedure. The instructor followed a scripted protocol to ensure consistent delivery of the information, participants were also briefed on the think-aloud protocol; (2) adjustment and calibration of the equipment. For the VR setting, participants were instructed to adjust the interpupillary distance (IPD) within the device to achieve optimal visual clarity. After this adjustment, they received instructions on how to navigate the virtual environment. In the RL setting, participants put on and adjusted a helmet equipped with a 180-degree lens camera; (3) The wayfinding experiment. The participants looked for three specific locations within the building. During this phase, an observer accompanied the participants, taking notes on their behavior and their comments; then, in the final phase (4), we used a questionnaire method. The participants evaluated their impression of the building and the wayfinding process, and they were also asked to self-assess their own orientation skills.

3 Sample

Seven males and twelve females participated in the study in all three experimental settings. Participants were of varied ages: five aged 20–30, seven 30–40, six between 40 and 50, and one older than 50. In the RL-only setting, there were five participants: (M:F, 3:2), in VR-only setting, seven people (M:F, 2:5), and seven people in the RL&VR setting (M:F, 2:5).

We controlled for the previous knowledge of the building, and four people had visited the building before; one person was unsure. For the RL&VR setting, we set a gap period of a minimum of six months between the RL and VR sessions. The height of participants ranged from 158 to193 cm. Also, wayfinding ability in new conditions was assessed on a scale from 1–very weak to 5–very strong ($M = 3.84$, $SD = 1.01$).

4 Results

We conducted route tracking and analyzed user behavior to compare RL and VR behavior. We divided the building into zones and then identified structural elements (e.g. corridor A, staircase, curved corridor)—see Fig. 1. Upon that, we could describe each route using steps; a step was defined as a part of the route in which the respondent chose to use a specific structural element (e.g. a specific corridor, elevator, information board). Using a predefined step allowed us to compare routes chosen by different participants and help find similarities and differences in real-life and virtual reality wayfinding strategies (see Fig. 1, presenting the steps for one of the destinations, room 2.20).

In both settings, participants began their routes by using the main stairs or elevator. However, in RL settings, almost all participants first directed themselves to the information board, whereas in VR settings, only two participants did so. Despite these differences, participants often followed similar structural elements to reach the target location, such as the mezzanine and the curved corridor. Less frequently used routes included the side corridors (Corridors D, or E).

Below, we present the specific results for room 2.20, but we calculated the average number of steps necessary to reach each of the experiment's destination targets (see Table 1).

In general, there were more steps in the VR setting, yet the difference was small, with 1–3 steps on average, and one exception for the route to room 1.2, in which the number of steps was higher in RL-only conditions than in VR-only. Also, in both conditions, the third task was completed in the fewest number of steps, due to learning during wayfinding.

In our analysis of the steps taken by participants to locate a room on the second floor, we focused on the challenge posed by the building's design, where the most visible main stairs only lead to the first floor. The data, presented in Fig. 2, highlight three common strategies observed across both conditions and experimental groups, as well as two less frequent strategies chosen by only a few participants.

The most common approaches included: using the elevator to go directly to the second floor (9 participants); taking the main stairs to the first floor and then using the elevator to reach the second floor (7 participants); taking the main stairs, the mezzanine area, and then corridor B that links the first and second floors (5 participants).

In contrast, five participants chose less conventional paths, using hidden staircases located in the building's narrowest areas. These participants attended the experiment in both conditions (RL & VR). These diverse strategies highlight the variety in decision-making processes among participants and demonstrate how spatial layout influences wayfinding behavior during the initial attempt to locate a destination. In the second attempt to reach the second floor, none of the participants used the hidden staircases, instead opting for faster and more direct routes.

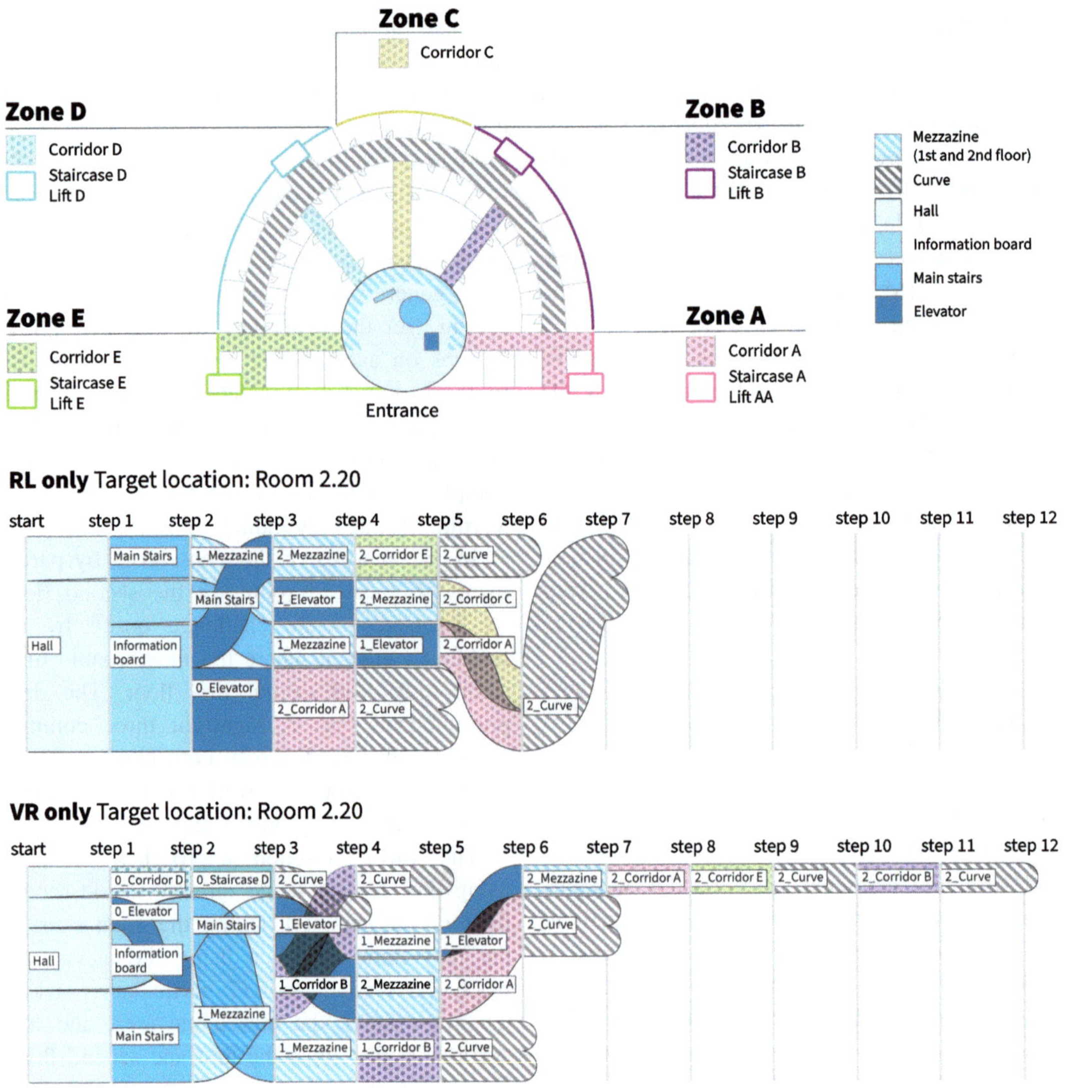

Fig. 1 Structural floor plans and steps visualisation for room 2.20

Table 1 Number of steps in each experimental setting for all routes

Route to	RL-only		VR-only		RL in RL&VR		VR in RL&VR	
	M	*SD*	*M*	*SD*	*M*	*SD*	*M*	*SD*
Room 2.20	3.32	3.76	6.71	3.04	4.94	3.52	7.32	3.29
Room 1.2	1.56	4.16	6.24	2.81	6.12	3.14	4.46	2.04
Room 2.42	2.56	3.29	5.92	2.64	3.81	3.11	5.7	2.58

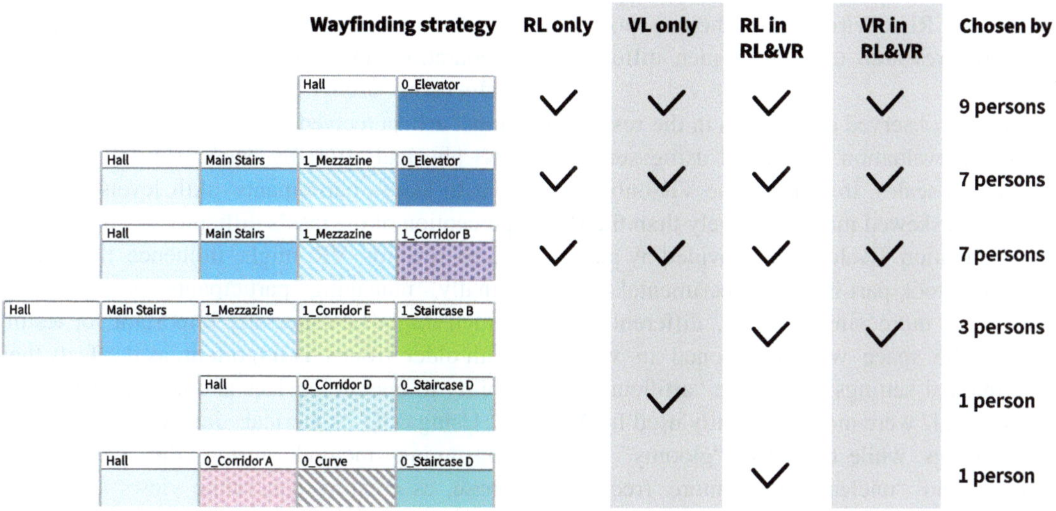

Fig. 2 Comparison of wayfinding strategies for the first attempt to reach the second floor

5 Discussion

In general, participants took more steps in the VR setting, yet the difference was small (1–3 steps on average, with one exception). However, in both settings, the number of steps declined as the participants learned the building's topology.

We also observed that patterns of behavior were similar across both settings. Participants often returned via different routes, and their choices of structural elements (e.g., stairs, elevator) were comparable in both RL and VR settings. Notably, the most visible and prominent elements in the space were more frequently chosen. This similarity was evident when examining wayfinding strategies to reach the second floor. Participants across various settings and groups employed several common approaches. These included straightforward methods, such as taking the elevator directly to the desired level, as well as more complex routes that combined multiple elements, such as using the main stairs. The main stairs, while prominent in the main hall, only reached the first floor, prompting some participants to switch to the elevator to access the second floor. A less common strategy involved navigating through the less obvious corridor B, which provides access to the second floor.

Additionally, some participants displayed more individualized and unconventional behaviors by choosing less visible or hidden staircases (located at the ends of specific corridors), such as staircases labeled D, or E (Fig. 2). These choices, while infrequent, suggest that certain individuals either possessed a different spatial awareness of the environment or preferred alternative routes they perceived as more efficient or suitable for their needs. This variety in wayfinding behavior highlights the diverse decision-making processes employed by participants when navigating to higher levels within the building, particularly noticeable in the group participating in both conditions (RL & VR). Strategy differences could also be attributed to the lack of wayfinding information, such as directional cues. As a result, participants' decisions were primarily based on their understanding of the space.

Some of the elements in the building were more easily noticed in the real-life setting than in VR. The only wayfinding element in the main hall (near the entrance) was an information board listing rooms and their functions, which was rarely used voluntarily in the VR setting. Another observed difference in behavior was the frequent choice of the elevator in the RL setting. Since these elements are experienced differently in VR and RL environments, there is potential for future research to explore such differences further.

We also observed differences in the results of the space evaluation conducted using semantic differential scales. In general, the VR-only evaluation was skewed more negatively than the RL-only evaluation. Evaluations provided by participants who took part in both experimental conditions were more similar. Also, different qualities of the space were mentioned in various experimental settings: traits like "artificial" and "impersonal" were more frequently used for VR-only settings, while traits like "gloomy", "confusing", and "unclear" were more frequently used for RL-only settings.

6 Conclusion

The results of our study confirmed that navigation behavior was similar across both real-life (RL) and virtual reality (VR) settings during the wayfinding process. Although respondents employed various strategies, their choices appeared to be influenced more by individual preferences than by the type of setting. These findings support the potential of VR as a valuable tool for wayfinding research, providing insights into how users develop spatial orientation and navigate buildings.

A notable positive outcome was that participants moved freely during the VR simulations and reported no discomfort. However, their impressions of the space differed between the two settings. Participants in the separate RL and VR groups described negative aspects of the environment differently, while feedback from the combined RL&VR group showed more alignment. Despite these differences, participants consistently perceived the virtual environment as somewhat artificial, which did not impact their wayfinding strategies. These results underscore the validity of VR for studying navigation while highlighting areas for further refinement in replicating spatial qualities.

There are also some practical remarks concerning the use of VR technology in wayfinding research. Regarding participants' preparation, clear instructions are crucial for influencing users' behavior as the VR conditions are perceived as less risky and require less physical effort by participants. It is important to assess participants' skill levels and their perception of the task's difficulty. Also, their attitudes towards VR might influence the results. Finally, managing participants' stress levels, which might be especially important for testing with older groups, is important, as it affects their ability to operate devices and accomplish tasks.

Using 3D spherical images is a time-consuming method during the development phase, as creating panoramic views and necessary corrections are essential. Yet, in the experimental phase, the speed and ease of conducting experiments are remarkable. Also, modifying views by editing, adding, or removing elements (e.g., information carriers, colors etc.) facilitates testing in early phases, for example when buildings are not yet erected. However, it should be noted that VR testing is not the best method for testing the legibility of typography—some visual information, including small typography, in our model required adjustments to maintain legibility from the same distance as in natural conditions (despite using the best resolution—8 K).

References

Al-Sharaa, A., Adam, M., Amer Nordin, A. S., Alhasan, A., Mundher, R., & Zaid, O. (2022). Enhancing Wayfinding performance in existing healthcare facilities using virtual reality environments to revise the distribution of way-showing devices. *Buildings, 12*(6), 790–808. https://doi.org/10.3390/buildings12060790

Dalton, R. C., Hölscher, C., & Montello, D. R. (2019). Wayfinding as a social activity. *Frontiers in Psychology, 10*, 142. https://doi.org/10.3389/fpsyg.2019.00142

Dong, W., Qin, T., Yang, T., Liao, H., Liu, B., Meng, L., & Liu, Y. (2022). Wayfinding behavior and spatial knowledge acquisition: Are they the same in virtual reality and in real-world environments? *Annals of the American Association of Geographers, 112*(1), 226–246. https://doi.org/10.1080/24694452.2021.1894088

Ewart, I. J., & Johnson, H. (2021). Virtual reality as a tool to investigate and predict occupant behaviour in the real world: The example of wayfinding. *Journal of Information Technology in Construction, 26*, 286–302. https://doi.org/10.36680/j.itcon.2021.016

Farr, A. C., Kleinschmidt, T., Yarlagadda, P., & Mengersen, K. (2012). Wayfinding: A simple concept, a complex process. *Transport Reviews, 32*(6), 715–743.

Feng, Y., Duives, D. C., & Hoogendoorn, S. P. (2022). Wayfinding behaviour in a multi-level building: A comparative study of HMD VR and desktop VR. *Advanced Engineering Informatics, 51*(101475), 1–22. https://doi.org/10.1016/j.aei.2021.101475

Goodroe, S., Velasco, P. F., Gahnstrom, C. J., Wiener, J., Coutrot, A., Hornberger, M., & Spiers, H. J. (2025). Predicting real-world navigation performance from a virtual navigation task in older adults. *PLoS One, 20*(1), e0317026. https://doi.org/10.1371/journal.pone.0317026

Goupil, V., Arnaldi, B., Argelaguet, F., Michaud, A.-S., & Gouranton, V. (2023, December 4). Can you find your way? Comparing Wayfinding behaviour between reality and virtual reality. In *ICAT-EGVE 2023—International conference on artificial reality and Telexistence & Eurographics Symposium on virtual environments*. Dublin.

Halik, Ł., & Kent, A. J. (2021). Measuring user preferences and behaviour in a topographic immersive virtual environment (TopoIVE) of 2D and 3D urban topographic data. *International Journal of Digital Earth, 14*(12), 1835–1867. https://doi.org/10.1080/17538947.2021.1984595

Haq, S., Hill, G., & Pramanik, A. (2009). Topological configuration in wayfinding and spatial cognition: A study with real and virtual buildings for design relevance. 1–8. https://www.semanticscholar.org/paper/f143e037e792981c2d2f100d5f34b1fa0f587306

Jamshidi, S., Ensafi, M., & Pati, D. (2020). Wayfinding in interior environments: An integrative review. *Frontiers in Psychology, 11*, 549628. https://doi.org/10.3389/fpsyg.2020.549628

Kuliga, S., Mavros, P., Brösamle, M., & Hölscher, C. (2020). Comparing human Wayfinding behavior between a real, existing building, a virtual replica, and two architectural redesigns. In J. Šķilters, N. S. Newcombe, & D. Uttal (Eds.), *Spatial cognition XII* (Vol. 12162, pp. 160–179). Springer International Publishing. https://doi.org/10.1007/978-3-030-57983-8_13

Lynch, K. (2005). *The image of the city*. The MIT Press.

Montello, D., & Sas, C. (2006). Human factors of Wayfinding in navigation. In W. Karwowski (Ed.), International encyclopedia of ergonomics and human factors , *Second Edition—3Volume Set* (pp. 1–17). CRC Press. doi:https://doi.org/10.1201/9780849375477.ch394.

Skorupka, A. (2009). Comparing human Wayfinding behavior in real and virtual environment. In D. Koch, L. Marcus, & J. Steen (Eds.), Proceedings of the 7th international space syntax symposium. (pp. 1–7) KTH.

Symonds, P., Brown, D. H. K., & Lo Iacono, V. (2017). Exploring an absent presence: Wayfinding as an embodied sociocultural experience. *Sociological Research Online, 22*(1), 5. https://doi.org/10.5153/sro.4185

van Beek, A., Duives, D. C., Feng, Y., & Hoogendoorn, S. P. (2024). Comparison of pedestrian wayfinding behavior between a real and a virtual multi-story building—A validation study. *Transportation Research Part C: Emerging Technologies, 163*, 104650. https://doi.org/10.1016/j.trc.2024.104650

Vilar, E., Rebelo, F., & Noriega, P. (2014). Indoor human Wayfinding performance using vertical and horizontal signage in virtual reality. *Human Factors and Ergonomics in Manufacturing and Service Industries, 24*(6), 601–615. https://doi.org/10.1002/hfm.20503

Yassin, M., El Antably, A., & Abou El-Ela, M. A. S. (2021). The others know the way: A study of the impact of co-presence on wayfinding decisions in an interior virtual environment. *Automation in Construction, 128*, 103782. https://doi.org/10.1016/j.autcon.2021.103782

Balance Between Functionality and Aesthetics. An Inclusive Design Standpoint

Francesco Burlando, Boyu Chen, and Niccolò Casiddu

Abstract

Corresponding the call of the European Green Deal to people's living spaces, the New European Bauhaus aims to facilitate the transformation of society along three major values: inclusion, sustainability, and aesthetics. However, the embracement of the latter in the design practices imposes for researchers a challenge and the objective of the paper is about exemplifying the balance between aesthetic components and functionality in inclusive design. This paper presents the evolution of different disability models, from medical model, social model to human development model, which improves the comprehension of disability, such as: the impact factors, and final scope ecc. Recognizing the role of aesthetics, the inclusive solutions correspond to the emotional dimension for reaching positive experiences and well-being of users. The analysis of three well-established cases illustrates the solutions to discussing how to reach an inclusive environment and be emotionally pleasing.

Keywords

Inclusive design · New European Bauhaus · Aesthetic experience · Urban Design · Vulnerable population

1 Introduction

The "New European Bauhaus" links the "European Green Deal" to people's living spaces and promotes social changes in the three values of inclusion, sustainability and aesthetics (Van Reusel et al., 2022). Integrating the latter poses a challenge for researchers and designers. Urban space is a fundamental aspect of an individual's well-being. Design is the ability of human beings to shape their own environments (Heskett, 2005): it must satisfy different functional needs as well as contribute to the aesthetic and lifestyle aspirations of the users and provide a greater sense of personal well-being. Inclusive design is a design perspective and methodology that explores the needs of disadvantaged and disabled people to integrate into mainstream society.

This paper will illustrate the balance between aesthetic components and functionality in inclusive design, which contributes in improving disadvantaged groups to live with autonomy, equality and dignity, while recognizing the diversity of people's needs.

F. Burlando (✉) · N. Casiddu
University of Genoa, Genoa, Italy
e-mail: francesco.burlando@unige.it

N. Casiddu
e-mail: casiddu@unige.it

B. Chen (✉)
University of Campania Luigi Vanvitelli, Caserta, Italy
e-mail: boyu.chen@unicampania.it

S. Camplone and C. Porfirione (eds.), *Design for All and the Future of the Planet*, Design For Inclusion 6, https://doi.org/10.1007/978-3-032-21769-1_4

2 Disability Models and Inclusive Design

There are many different definitions of disability, so we started by analyzing the disability models to better understand these concepts. In fact, a disability model is an invaluable analytical framework that enables researchers to gain a more comprehensive understanding of the nature of disability, cause and effect.

2.1 Different Disability Models

2.1.1 From Medical Model to Social Model

As for the medical model, disability is a problem of impaired function and structure of an individual's body caused by disease, injury or other conditions. It requires only medical diagnosis and treatment, without regard to social and environmental factors. People with disabilities are viewed as "patients" who deviate from normalcy and are viewed in a passive and negative manner, even if society and the environment are considered to be responsible for the situation of people with disabilities. In terms of design, interventions are made only with aids (e.g., prosthetic limbs, wheelchairs, etc.), without additional help at the social and urban levels, and without realizing the vision of an individual's self-determined life. However, the affirmative action movement triggered a broad discussion on the rights of people with disabilities, attracting the attention of the academic community and making an impact. In the process, relevant sociological scholars have attempted to articulate and study this from a social context. Nagi (1965), for instance, sees disability as a social manifestation of physical or mental limitations and a mismatch between an individual's abilities and the demands of their own and the social environment. He also noted the dynamic nature of disability and the role of the social environment in the disability process. The social model of disability was developed by Oliver (1996), who was unequivocal in his assertion that disability arises from within society and that society fails to meet the needs of people with disabilities. Therefore, the problems faced by people with disabilities are socially constructed and shaped. The World Health Organization (WHO), as a global health organization, has changed its understanding of disability from the 1980s publication of International Classification of Impairments, Disabilities, and Handicaps (ICIDH) to a more generalized understanding of disability. Based on that, WHO (2001) proposed an adjusted version, International Classification of Functioning Disability and Health (ICF). The ICF framework defines disability as a combination of physical, personal and social dimensions that are in a state of dynamic change. The emergence of new ideas has changed the design thinking under the traditional healthcare model. It has started to focus on and tackled environmental factors. Additionally, it's considered that people are often labeled as disabled due to poor design, poor services and environments, and cultural biases (Clarkson & Coleman, 2015).

2.1.2 Human Development Model

However, the understanding of disability did not end with the previous theories. The social model has been criticized for ignoring the fact of physical impairment. According to Yang's (2015) interpretation, the social model ignores the physical condition of the disabled subject, resulting in them being viewed as "half-humans" with only social attributes. This is a flawed approach that needs to be addressed. The ICF model, developed from the social model, fails to adequately address the development and well-being of people with disabilities and requires further optimisation. In the context of emphasizing the human dignity of people with disabilities, there is a clear call for a new human rights model. Mitra (2018) optimizes the social model to construct a human development model for people with disabilities, defining the disability as a group of people who have a lack of capability or functioning, such as a simple impairment not possessing the previous situation.

Within the framework, capability/functioning is positioned in a crucial place. Capability refers to the actual opportunities and freedom of choice that can be accessed in a given setting, for instance, that an individual holds the ability to have access to education, even though they may ultimately choose not to continue their education, is an ability or liberty in itself. In addition, functioning refers to achievement or the realization of a capability, such as, obtaining a relevant degree is a clear indication that an individual is able to pursue higher education. Both are directly linked to the development and well-being of the individual. Resources, represent the goods, services, etc. that an individual has access to, encompassing the potential for human well-being that an individual can have. Structural factors refer to the external environment that affects the individual. With all aspects mentioned, the ICF model considers the health of the individual as the ultimate goal, and targets the development and well-being of people with disabilities and emphasizes the differences and agency of the individual. To match the goal, the creations of design must bridge the gap between life and the individual, which enables users to utilize their self-motivation and develop themselves and live more positively.

2.2 Inclusive Design

Inclusive design is the design of mainstream products and/or services that are accessible to, and usable by, as many people as reasonably possible, without the need for special adaptation or specialised design (Normie, 2005). Inclusive design values equality and respect for user diversity. It benefits everyone. Updating the disability model will make inclusive design more relevant to real life.

Inclusive design is a well-established concept that has been supported by a vast body of academic literature for decades.

However, aesthetic considerations are less readily incorporated into standardized categories and have not always been the focus of design throughout history. Existing research on inclusivity is lacking in psychological and emotional perspectives of users. Inclusive solutions are often unattractive and sometimes lack inclusive aesthetics. Even if they are functionally inclusive, they do not appeal to the emotions of people with disabilities or disadvantaged groups (Patrick & Hollenbeck, 2021). Emotions affect the user's experience and expectations for the next time they use the product, and this directly impacts their quality of life (Cooley et al., 2019).

Inclusive design is based on a clear conceptual understanding of different disabilities, a dynamic approach to disability, and the agency of people with disabilities. Microsoft has proposed design principles for understanding disability and change (Shum et al., 2016). These principles recognise that everyone at some point in their lives has a temporary or permanent reduction or loss of individual function. Steinfeld and Maisel set out eight goals that not only require inclusion on a functional level, but also emphasize understanding on a social and cultural level to meet emotional needs (Steinfeld & Maisel, 2012). This helps users to live a more active life with dignity. The following paragraphs will analyze functional and aesthetics aspects which point out that good spatial and environmental designs need to satisfy both and be achieved simultaneously.

3 Case Study

Designers and researchers need to consider both aesthetic appropriateness, focusing on the psychological needs of vulnerable populations and exploring the inclusion of emotional content in products, and cultural sensitivity, understanding how knowledge impacts cognitive and usage behaviors influenced by deep cultural factors (Zhao, 2022). Meeting these criteria will better enable individuals with disabilities to achieve Mitra's vision of an independent life. Therefore, this section will analyze three cases that exemplify effective integration of aesthetic elements into projects without compromising functionality.

Case 1 Yinian Rooftop Garden, Chongqing, China

The global trend of aging is a fact that cannot be ignored. It is clear that significant increases in the elderly population are occurring in most developed and some developing countries. The Yinian Rooftop Garden for Seniors project in Chongqing, China, provides a suitable culturally aging environment for elderly residents in the center of Chongqing under the guidelines of inclusive design and people-oriented strategy.

Aging in place as an appropriate approach exemplifies well how to be people-oriented in the project. Innovating the traditional way of aging, Aging in place is gaining attention, which involves not leaving the familiar environment of the community and going to a separate care facility. The elderly care more about interpersonal communication in the caring facilities, rather than the caring service (Barrett et al., 2012). In addition, comprehending Chinese cultural values, derived from Confucian philosophy, are constructed by interpersonal relationships and social orientation, and include the orientation between man and nature. So the Chinese believe that man is a part of nature and must learn how to adapt to nature in order to achieve harmony.

In this case, the facilities are located at the existing residential building, which enables the elderly to maintain the previous interpersonal networks and living routine, without at the expense of the social connections for aging. Also the project explores and reutilized the unused roof as an recreational space, in which it's designed the activities with green plants for corresponding the habits of the elderly, originated from Chinese culture. Additionally, the project studied the physiological characteristics and needs of elderly people in different health situations, such as elderly who have weak limb strength and balance or may be in wheelchairs. The project provides more safety measures to avoid bumps and bruises in the process of activities and the safety system runs through the whole space and every detail. Specifically, 750 mm/900 mm of dobble-level of handrail, and 2400 mm spacious aisles are embedded in the spaces consenting the passages for 2 wheelchairs contemporaneously. In addition, it lowers the height of the flower beds at 650 mm to accommodate the elderly in wheelchairs, allowing them to perform accessiblely gardening operations with friends or family.

Addressing the issue of aging, the Yinian Rooftop Garden exemplifies well the combination of respecting the emotional needs of the Chinese elderly and not sacrificing the functionality. It contributes to this project to improve the well-being of the elderly and support them to age more actively.

Case 2 Friendship Park, Montevideo, Uruguay

Urban parks are an essential part of urban life. They are not only places for social and cultural activities, but also key spaces for enhancing the quality of life of residents. Parks must be designed and equipped to take into account the holistic development of children, including their physical and mental health (Tian et al., 2022). The following project, Friendship Park, will illustrate how to design the public park with high-quality inclusion.

Friendship Park, a park project located in Montevideo, Uruguay, is designed by architects Marcelo Roux and Gastón Cuña to provide a barrier-free recreational and learning environment for children and adolescents of diverse ages and ability levels. Multi-sensory is an inclusive approach that engages with the visual, tactile, auditory, olfactory, and tasting domains, and provides the possibility for the people with diverse capabilities to interact with the products and spaces (Altay, 2017), functioning well also for sensory-impaired children reducing negative emotional experiences. A multi-sensory approach to participation is the most engaging way to promote positive social behavior (Hussein et al., 2016).

The park is designed with careful multi-sensory considerations that create an accessible and pleasant environment to stimulate children's curiosities and desires to explore through rich sensory experiences. To reach the goal, a level difference between the park and the avenue for creating a safe space and reducing the noises. The park also transforms the original sloping plane into a flat horizontal plane to ensure an accessible environment. The pavement is widened and leveled to match the specific needs of the children, and diverse play facilities with large extension of measures are added for the children in wheelchairs. Additionally, the park also prioritizes aesthetic experience, ensuring that the park is not only functional but also emotionally appealing. A garden with a variety of plant species and materials corresponding to the multi-sensory needs, provides color, scent and touch for enriching the playing experiences. The colorful gardens and diverse play facilities provide children with enjoyable sensory pleasures and inclusive interactive experiences. The scent of plants also can facilitate children with sensory impairments to utilize their sense of smell to perceive their environment, positively influence their mood and health, and be integrated in the crowd equally.

This park not only embodies the principle of inclusivity, but also promotes children's social behavior and emotional development, providing them with a place to engage equally with their peers and pave the way to reach their full potential.

Case 3 MAC house, Thiene, Italy

The personalization offers the potential for users having the expression of personal preferences (Steinfeld & Maisel, 2012). Through the innovative choice of materials and space planning, a living environment that is easy to navigate and adapt to is tailored for a blind lady at Thiene, Italy.

The capability of living autonomously for a population suffering from a visually impaired situation is vital in their daily life. The clear boards of the environment facilitate them to comprehend the spaces and increase their sense of control that alleviate the anxiety or loss (Mediastika et al., 2020). Human access to information about the spatial environment relies on the synergy of sight, hearing, touch, smell and multiple senses. When one of the senses is weakened or absent, the sensitivity of the other senses will gradually increase as a complement to acquire the information in the spaces (Kristjánsson et al., 2016). The floor is pavemented by the stone and porcelain, utilizing the tactile compensation that addresses the choices of simple glyphic language, which is embedded by the clear textures on the materials' surface. The obvious and simple oriental system is around a singular corridor spine, with the aim to reduce the maze effect and ensure efficient movement. Different daily use or activity, as nodes, are connected and settled to align to the corridor spine, following and facilitating the learning process of the client. Additionally, the consideration of sunlight also helps and enriches navigational activities.

Deviating from the traditional ways, the inclusive approaches in this project combine the functional and emotional dimensions dynamically, with the profond comprehension of physiological characteristics and emotional preferences of the client.

4 Conclusion

The three case studies cited above illustrate how achieving functionality and aesthetics is not a dualistic issue. Although complicated to achieve, the goal of making the two coexist is essential to enable disadvantaged user groups to truly enjoy equal dignity. In the case studies shown, this goal is achieved with different approaches and with reference to different groups of disadvantaged users, yet always respecting the guidelines of Inclusive Design. These virtuous examples set the challenge to designers for the future, emphasizing once again how a good design is one in which the inclusive component is so well integrated that it is not even noticeable, not only from a functional point of view, but also aesthetically. In the future, we advocate that this paper can serve as a trigger for more attention on the relative topics, and contribute to a society with more inclusive common sense.

Acknowledgments The paper is the result of a joint reflection by the Authors carried out with the supervision of Niccolò Casiddu. The "Introduction" and "Disability models and inclusive design" paragraphs have to be attributed to Boyu Chen, "Case study" and "Conclusion" to Francesco Burlando.

References

Altay, B. (2017). Multisensory inclusive design education: A 3D experience. *The Design Journal, 20*(6), 821–846.

Barrett, P., Hale, B., & Gauld, R. (2012). Social inclusion through ageing-in-place with care? *Ageing and Society, 32*(3), 361–378.

Clarkson, P. J., & Coleman, R. (2015). History of inclusive design in the UK. *Applied Ergonomics, 46*, 235–247.

Cooley, S., Burkholder, A. R., & Killen, M. (2019). Social inclusion and exclusion in same-race and interracial peer encounters. *Developmental Psychology, 55*(11), 2440.

Heskett, J. (2005). *Design: A very short introduction* (Vol. 136). Oxford University Press.

Hussein, H., Omar, Z., & Ishak, S. A. (2016). Sensory garden for an inclusive society. *Asian Journal of Behavioural Studies, 1*(4), 33–43.

Kristjánsson, Á., Moldoveanu, A., Jóhannesson, Ó. I., Balan, O., Spagnol, S., Valgeirsdóttir, V. V., & Unnthorsson, R. (2016). Designing sensory-substitution devices: Principles, pitfalls and potential 1. *Restorative Neurology and Neuroscience, 34*(5), 769–787.

Mediastika, C. E., Sudarsono, A. S., Kristanto, L., Tanuwidjaja, G., Sunaryo, R. G., & Damayanti, R. (2020). Appraising the sonic environment of urban parks using the soundscape dimension of visually impaired people. *International Journal of Urban Sciences, 24*(2), 216–241.

Mitra, S. (2018). The Human Development Model of Disability, Health and Wellbeing. In: *Disability, Health and Human Development* (pp. 9–32). Palgrave Macmillan, New York.

Nagi, S. (1965). Some conceptual issues in disability and rehabilitation. *Sociology. rehabilitation.*

Normie, L. R. (2005). BS 7000–6: 2005 design management systems–part 6: Managing inclusive design, by British standards 2005.

Oliver, M. (1996). Understanding disability: From theory to practice.

Patrick, V. M., & Hollenbeck, C. R. (2021). Designing for all: Consumer response to inclusive design. *Journal of Consumer Psychology, 31*(2), 360–381.

Shum, A., Holmes, K., Woolery, K., Price, M., Kim, D., Dvorkina, E., & Malekzadeh, S. (2016). Inclusive: A Microsoft design toolkit. *M. Price etc. Microsoft Des, 32*, 32.

Steinfeld, E., & Maisel, J. (2012). *Universal design: Creating inclusive environments.* John Wiley & Sons.

Tian, J., Zhang, Q., Li, J., & Zhang, B. (2022). Research on spontaneous play and environmental characteristics related to Children's interaction with plants in community parks of Tianjin. *Landscape Architecture Frontiers, 10*(3), 38–65.

Van Reusel, H., Boutsen, D., & Barosio, M. (2022, August). The new European Bauhaus in architecture education? An anthology for the invisible. In *Annual conference of the European Association for Architectural Education* (pp. 128–143). Springer Nature Switzerland.

World Health Organization. (2001). *International classification of functioning.* WHO. www.who.int/classifications/icf/en/

Yang, Z. (2015). Institutions and life for people with disability: From 'individual model'to 'universal model'. *Chinese Journal of Sociology, 35*(6), 85–115.

Zhao, C. (2022). Building the theoretical model for aging design: Intuitive cognition, prior knowledge and familiarity in the context of technology and culture. *Zhuangshi*, (05), 12–19. https://doi.org/10.16272/j.cnki.cn11-1392/j.2022.05.001

User Integration Funnel: A Theoretical Model for User Research Practices

Annapaola Vacanti

Abstract

The paper introduces the User Integration Funnel, a theoretical model for enhancing user participation in the design process, within the framework of Human-Centered Design (HCD) and Human-Technology Interaction (HTI). As the integration of data into design becomes increasingly prevalent, the model responds to the challenges of balancing quantitative insights from big data with the nuanced, contextual understanding provided by thick data. This dual approach emphasizes inclusivity and adaptability, fostering designs that not only function effectively but also resonate with diverse user experiences. The model conceptualizes user engagement through three concentric levels: sporadic involvement, sustained observation, and active collaboration. Each level explores a distinct interaction type, knowledge accessibility, and output, ranging from user opinions to behavioral insights and co-created ideas. These levels align with mixed-method research paradigms, blending qualitative and quantitative methods to address the complexities of HTI projects. According to a systemic perspective, users are situated within interconnected human, technological, and ecological networks. HCD theory must acknowledge practical challenges, such as power dynamics, resource constraints, and varying levels of user literacy, advocating for progressive engagement to ensure meaningful participation. While the framework promotes inclusive, adaptive technologies, it raises critical questions about the implications of hyper-adaptivity. By minimizing friction and resistance in human experiences, design risks undermining the critical thinking, creativity, and resilience that emerge from overcoming challenges. It is thus urgent for designers to balance the benefits of seamless interaction with the preservation of human autonomy, creativity, and growth, through a holistic, humane approach to the co-evolution of humans and technology.

Keywords

User research · Human technology interaction · Co-design · User participation · Humane technology

1 Human—Technology Interaction and Designing with Data

Since the dawn of humanity, artificial solutions have played a critical role in reducing effort, simplifying tasks, and enabling achievements beyond physical and cognitive limitations. Collectively referred to as *technology*, these

A. Vacanti (✉)
Università Iuav di Venezia, Venice, Italy
e-mail: avacanti@iuav.it

S. Camplone and C. Porfirione (eds.), *Design for All and the Future of the Planet*, Design For Inclusion 6, https://doi.org/10.1007/978-3-032-21769-1_5

artifacts have evolved into complex, interconnected, and often autonomous systems, bridging the physical and digital realms. The design discipline, broadly defined by Buchanan as the conception and planning of the artificial (Buchanan, 1992), is pivotal in determining the purpose, functionality, and usability of these non-natural artifacts.

In today's era, our tools not only assist us but also collect and record behavioral data, generating an immense volume of information that transforms human lives into data streams, a phenomenon known as *datafication* (Cukier & Mayer-Schoenberger, 2014). This trend generates valuable insights but also raises concerns about prioritizing technological interests over human needs (Galimberti, 2023). As technological advancement accelerates, humanity's relationship with technology is shifting from coexistence to *co-evolution*, marking the dawn of a *hybrid age* characterized by ubiquitous, intelligent, and social technologies (Khanna & Khanna, 2013).

However, not all individuals can harness the full potential of available technology. The *Technology Quotient* (TQ) (Nosta, 2017) quantifies an individual's ability to adapt and integrate technology, yet global technological literacy remains alarmingly low. Reports from the OECD (2016) reveal that in 33 industrialized countries, only 5% of the population possesses advanced technological skills, with the majority struggling to complete moderately complex tasks. This discrepancy underscores a profound gap between the competencies of average users and the so-called technological elite, including engineers and designers (Nielsen, 2016). The reliance on vast computational capabilities and global databases fosters a quasi-mystical faith in data and algorithms (Davis, 2015), often sidelining critical human oversight (Anderson, 2008). Data-driven processes frequently reflect inherent biases in their design and datasets, emphasizing the need for human critique in interpreting information meaningfully. Human - Technology Interaction (HTI) thus stands as an iterative discipline that hybridizes technical knowledge with humanistic insights, gathered by involving users in the design process.

In the context of data, the growing reliance on big data marks a profound transformation in design. Characterized by an abundance of data, this trend enables unprecedented optimization but often sacrifices inclusivity by marginalizing minorities and reinforcing biases (Criado Perez, 2019). Conversely, *thick* data (Wang, 2016), derived from qualitative methods from small datasets, provides nuanced insights into user contexts and emotions, complementing the breadth of big data. The integration of both approaches creates a holistic understanding essential for inclusive and human-centered design.

For example, while big data offers scalable insights into user behavior through algorithms and statistical models, it often lacks the depth required to understand unique user needs or cultural contexts. *Thick* data, collected through interviews, focus groups, and ethnographic methods, fills this gap by capturing emotional and social dimensions. Together, these methodologies enable designers to balance scale with depth, fostering innovative and inclusive solutions (Fig. 1).

Additionally, emerging frameworks such as *data-informed design* (King et al., 2017) and *data-enabled design* (Bogers et al., 2016) offer structured approaches for integrating data into the design process. These methodologies empha-

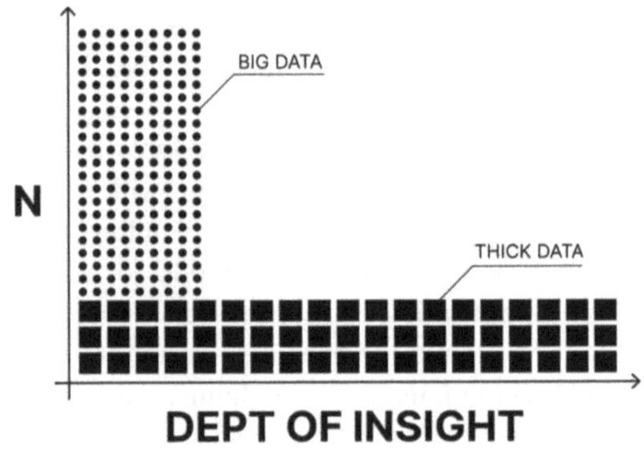

Fig. 1 A comparison between big data and thick data. Where thick data loses in scale, big data loses in resolution (credits of the author, 2024)

size collaboration between designers, users, and data scientists, promoting iterative cycles of prototyping and testing. The hybridization of big and *thick* data has also led to frameworks like *big-thick blending* (Bornakke & Due, 2018), which calibrates and contextualizes large-scale data insights with qualitative observations to add depth and understanding.

As the role of data in design evolves, designers must adopt new competencies to navigate the challenges of a data-driven world. They assume roles as translators, team leaders, and analysts, bridging gaps between technical and human-centered perspectives as categorized by Lu et al. (2021). For instance:

1. Designers as *translators* facilitate communication between disciplines, requiring technical literacy and strong interpersonal skills;
2. Designers as *empathizers* leverage their ability to interpret user needs, even in the context of large datasets;
3. Designers as *team leaders* guide projects by combining data insights with intuitive decision-making;
4. Designers as *analysts* assess data for meaningful patterns, ensuring designs remain relevant and user-focused.

These roles highlight the growing complexity and interdisciplinarity of the design profession. As translators, designers bridge technical and human dimensions, fostering collaboration among diverse stakeholders and aligning technical frameworks with cultural and social contexts. As empathizers, they humanize data-driven processes, addressing big data's limitations by integrating qualitative insights to create trustful, user-centered solutions. Team leaders balance data-driven rigor with creative intuition, mediating conflicts and aligning multidisciplinary goals. As analysts, designers critically evaluate data, identifying biases and contextualizing findings to ensure designs are not only functional but also equitable and inclusive.

Hybrid methodologies that balance quantitative insights with qualitative depth are vital for addressing the growing complexity of user needs. By integrating these approaches, designers can ensure that data serves as a tool for understanding, rather than dictating, human experiences. This paper aims to contribute to this dialogue by introducing the User Integration Funnel, a theoretical model that enhances user participation in design processes. Through its framework, the model demonstrates how different levels of interaction, knowledge accessibility, and collaboration can foster inclusivity and enable human-centered technological innovation.

2 User Participation Theory

The role of the designer lies in applying his/her expertise proactively to the existing, producing either incremental or even disruptive innovations (Giacomin, 2014). This creative synthesis activity is rooted in interpreting the complex factors driving societal change, aiming to deliver solutions that address people's needs, expectations, and desires, potentially even proposing new lifestyles (Tosi, 2018). Within this framework, Human-Centered Design (HCD) has emerged as a distinct approach, prioritizing people over the creative and aesthetic processes of designers or the technical specifications of the artifacts themselves. Originating in the 1970s under the term User-Centered Design (UCD), it was redefined as HCD and later expanded in *Humanity Centered Design* by Norman (2023), emphasizing the focus on people rather than their roles as end-users.

The HCD approach finds parallels in ergonomics, sharing its concern for human well-being while also extending into areas such as HTI and User Experience (UX). These fields focus on optimizing the global experience of interacting with an artifact, delving into the desires and expectations of users. The HCD process focuses on increased effectiveness, efficiency, and user satisfaction while mitigating potential adverse effects on health, safety, and performance. This iterative and multidisciplinary approach ensures that the user needs guide design, involving users throughout the process and focusing on the entire user experience.

However, the practice of HCD is not without challenges. Designers and organizations may underestimate its value, failing to account for systemic issues or creating artifacts that are inflexible and narrowly tailored. Such shortcomings are exemplified by the *Five Fundamental Fallacies* (Pheasant & Haslegrave, 2005), which include misconceptions such as designing for the average user or over-relying on personal intuition. Rubin and Chisnell (2008) stress the iterative nature of HCD as an empirical process, grounded in understanding human diversity to adapt designs to behaviors rather than expecting users to conform to pre-existing solutions. As technology becomes more intelligent and context-sensitive, especially in domains such as IoT and assistive robotics, emotional and psychological factors increasingly shape user acceptance, demanding nuanced approaches to design that balance usability with personal and contextual adaptability.

The question of how to effectively engage users in the design process has sparked critical debates. While user involvement is fundamental to HCD, it can inadvertently constrain innovation if users are unable to articulate future needs or rely on familiar concepts (Van Kleef et al., 2005). The dynamic between user insights and designer expertise necessitates a spectrum of methods for capturing explicit, observable, and latent user knowledge (Sanders, 2005). These include interviews, ethnographic studies, and co-design workshops, each offering distinct advantages and challenges depending on the phase and focus of the design process. The interplay between these methods highlights the need for flexibility and adaptability in how designers approach user research.

Central to HCD is the acknowledgment of users as carriers of domain-specific knowledge, with designers acting as facilitators who translate this knowledge into actionable insights. This shift in roles reflects broader disciplinary trends, where participatory and co-design approaches encourage users to become active contributors. These methodologies, rooted in the Scandinavian tradition of democratizing design, aim to empower users by leveraging their tacit knowledge while fostering mutual learning between stakeholders (Spinuzzi, 2005). Co-design, in particular, emphasizes creating shared design languages and tools, enabling collaborative ideation and prototyping that transcend traditional hierarchies.

Despite the strengths of HCD, contemporary design challenges demand a broader perspective that accounts for the systemic and ecological implications of design decisions. Designers must move beyond user-centricity to consider the complex interplay of human and non-human actors within interactive systems (Norman, 2023; Slavin, 2016). This systemic approach requires balancing empathy for individual users with an awareness of the broader socio-environmental context, acknowledging that design interventions often have ripple effects extending far beyond the immediate user base.

The convergence of various design disciplines toward shared reflections on the human-technology relationship underscores the evolving nature of design as a field. From ergonomics to UX, interaction design to data-driven methodologies, these domains increasingly intersect in their focus on creating meaningful, inclusive, and context-aware experiences. The proliferation of Living Laboratories exemplifies this interdisciplinary synthesis, merging HCD principles with data-driven insights to foster open, participatory innovation. By situating research within real-life contexts, Living Labs bridge the gap between controlled experimentation and the chaotic realities of everyday life, yielding insights that inform more robust and adaptable design solutions.

The trajectory of HCD reflects a shift from addressing individual needs to navigating the complexities of interconnected systems. As Van Mensvoort (2022) suggests, the future of design lies in embracing a humane approach—one that prioritizes compassion, inclusivity, and the co-evolution of humans and technology within a shared ecological framework. This holistic perspective challenges designers to redefine their role, not as isolated problem-solvers but as facilitators of collective innovation within a diverse and dynamic world. The synthesis of these reflections forms the foundation for advancing the discipline, fostering a more integrated and responsive design practice.

The User Integration Funnel.

User research practices constitute a vast and diverse toolbox available to researchers and designers to inform their work and make effective, informed choices. Data collection activities are increasingly becoming an integral part of the design process, both in academia and industry, making it essential for designers to engage with their users (Rohrer, 2014). Various methods allow for leveraging users' knowledge on a given topic at different levels, either by listening to them, observing their behaviors, or enabling them to propose (or co-create) innovative solutions (Westerlund, 2007). In this perspective, users are considered key actors in placing their daily life ecosystems at the center of the ideation, experimentation, and evaluation of technological artifacts. However, existing methods often face challenges that risk undermining, or at least diminishing, the reliability of their outcomes. Quantitative approaches have the advantage of enabling the collection of large datasets at relatively low costs, which can also be made accessible through open access for the scientific community. The study of such aggregated data offers a precise understanding of observable behaviors (Veeckman & Van Der Graaf, 2015). However, in the context of HTI, quantitative methods fall short in at least two aspects identified by Romero Herrera (2017): exploring the appropriation of technology and the adoption of new practices by users within their life contexts, and involving users in the ideation and evaluation phases of artifacts.

Qualitative methods seem preferable in addressing these gaps, although they also present challenges. For instance, co-design or usability testing activities often require users to replace mature technologies with unstable or incomplete solutions, which significantly impacts their assessment of the potential adoption of such products (Ogonowski et al., 2013). Researchers must also account for the cognitive effort and time commitment required from participants, who may easily lose motivation and interest, leading to feelings of burden, annoyance, or boredom (Rek et al., 2013).

Creswell and Clark (2011) coined the term *Mixed Methods Research* (MMR) to describe an approach that integrates both quantitative and qualitative methods to address research questions at various stages of the process. This integration can occur through three different approaches:

1. Connecting: a qualitative data source builds upon or follows a quantitative one (or vice versa);
2. Merging: data are collected using both quantitative and qualitative methods and then compared or interrelated;
3. Embedding: a qualitative dataset is embedded within a quantitative one (or vice versa) to explain one dataset's findings through the other.

These three MMR approaches illustrate the flexibility and adaptability of combining qualitative and quantitative data to address complex research questions. The connecting approach allows researchers to build on the strengths of one method to enrich or validate findings from the other, ensuring a comprehensive understanding of the subject. The merging approach fosters a more integrative perspective by comparing and interrelating datasets, revealing patterns or insights that may not emerge from a single method. Finally, the embedding approach is particularly valuable in providing deeper context to quantitative findings or validating qualitative observations through measurable data.

Designing technological solutions that are inclusive, acceptable, and humane in the sense proposed by Van Mensvoort (2022) requires leveraging human values and insights while empowering users to improve their conditions and enabling interactions that feel natural and resonate with their senses.

A visual metaphor of a funnel viewed from above illustrates three levels of user integration in the design process (Fig. 2), inspired by the

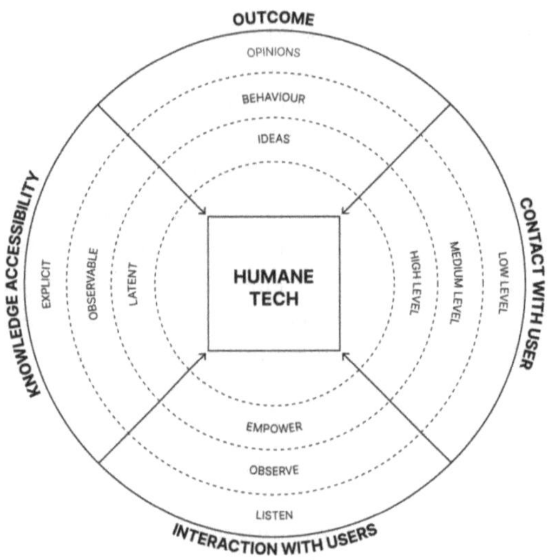

Fig. 2 The visual representation of the User Integration Funnel (credits of the author, 2024)

model for sustainable innovation in Living Labs (Romero Herrera, 2017).

The four quadrants of the diagram represent different types of interaction with users, the accessibility level of the knowledge sought, the degree of contact between researchers and users, and the outputs generated. The type of interaction defines various roles for users as contributors to design research.

- The outer ring represents sporadic involvement, where users share their opinions through activities guided by the designer.
- The middle ring indicates a more interactive role, where users act and interact within the study context, while their behaviors are observed and interpreted by the designer.
- The inner ring signifies active collaboration, where users are enabled to propose their ideas and evaluate solutions.

The level of knowledge accessibility reflects the type of user expertise being engaged:

- The outer ring pertains to explicit knowledge, which users articulate verbally.
- The middle ring relates to users' actions and observable skills in using a system.
- The inner ring involves users' creative abilities, which are latent and require activation through collaboration with designers.

The degree of contact describes the interaction intensity between users and designers:

- The outer ring represents brief engagements, such as interviews or similar activities.
- The middle ring involves sustained collaboration over time, requiring direct interaction during the observation period.
- The inner ring represents deep collaboration, fostering a close working relationship.

Outputs vary based on these parameters:

- The outer ring produces explicit opinions expressed by users.
- The middle ring generates an understanding of users' observed behaviors.
- The inner ring yields new ideas and insights applicable to the project.

The first level of integration is characterized by merging different data collection techniques. Quantitative data from surveys, self-reported diaries, or other methods are combined with qualitative data from interviews and focus groups. The second level uses connected data, linking information gathered through both methods to capture everyday practices empathetically and objectively. The third level involves connecting collaborative activity results with qualitative (e.g., self-reports, interviews) and quantitative methods (e.g., sensor recordings) to enrich the data. These data can provoke further reflections and insights among users.

An ideal innovation process encompasses all integration levels to address HTI project complexity comprehensively. Realistically, the appropriate integration level for each project depends on factors such as its scope and objectives, available resources, technical capabilities, timelines, and the expertise and availability of the designers, researchers, and users involved.

The funnel also incorporates a systemic perspective, recognizing the interconnectedness of human, technological, and ecological systems. By situating users within broader frameworks, the model aligns with the Gaia hypothesis, which envisions the Earth as a self-regulating system where humans are one of many interdependent components (Lovelock, 2000). This systemic thinking is essential for addressing the complex challenges of contemporary design, including sustainability, inclusivity, and ethical responsibility.

It is relevant to note that factors such as power dynamics, resource constraints, and varying levels of user literacy can pose barriers to effective collaboration (Holmes, 2020). Designers must navigate these obstacles by fostering trust, ensuring accessibility, and adapting their methodologies to the unique characteristics of their user base. By promoting progressive engagement, the funnel provides a pathway for overcoming these challenges, ensuring that user participation remains meaningful and impactful.

3 Conclusions and Openings

The discussions presented in this paper reflect on the complex interplay between humans and technology, exploring how design methodologies can mediate this relationship. The integration of user participation into the design process, organized through the User Integration Funnel, requires a balanced approach that leverages both data-driven insights and empathetic understanding. While the framework offers pathways to more inclusive and adaptive technologies, it also raises broader questions about the implications of designing for seamless interaction.

The contemporary design landscape increasingly prioritizes accessibility, inclusivity, and personalization, striving to create technologies that respond flexibly and intuitively to human needs. However, this trend toward hyper-adaptivity prompts critical reflection on the consequences of removing friction from human experiences. By shaping a world that offers less resistance and anticipates our desires, we risk creating environments that diminish the very challenges that foster critical thinking, creativity, and resilience (Vacanti, 2024). The act of overcoming obstacles, of engaging with complexity, is not merely a source of difficulty—it is a fundamental aspect of human development.

This reflection highlights a key tension in the designer's role. On the one hand, designers are tasked with improving usability, reducing cognitive load, and enhancing the quality of life through technology. On the other, they bear a responsibility to question whether total optimization serves human agency. Can we, as designers, strike a balance between enabling ease and preserving the capacity for struggle, learning, and growth?

Moreover, as technologies increasingly interconnect, the role of the designer must expand to consider the broader ecosystems in which these artifacts operate. Designing for human-technology interaction is no longer about creating isolated tools but about shaping systems where human and non-human actors coexist and co-evolve. This requires not only technical proficiency but also ethical and systemic awareness.

In conclusion, the design of inclusive, adaptive technologies must go hand in hand with a deeper consideration of their societal and psychological impacts. As we push the boundaries of personalization and seamless interaction, we must remain vigilant about preserving the essential human qualities of autonomy, creativity, and critical thought.

References

Anderson, C. (2008). The end of theory: The data deluge makes the scientific method obsolete. *Wired Magazine, 16*(7), 16–07.

Bogers, S., Frens, J., Van Kollenburg, J., Deckers, E., & Hummels, C. (2016). Connected baby bottle: A design case study towards a framework for data-enabled design. In *DIS 2016 - proceedings of the 2016 ACM conference on designing interactive systems* (pp. 301–311). ACM.

Bornakke, T., & Due, B. L. (2018). Big–thick blending: A method for mixing analytical insights from big and thick data sources. *Big Data & Society, 5*(1).

Buchanan, R. (1992). Wicked problems in design thinking. *Design Issues, 8*(2), 5–21.
Creswell, J. W., & Clark, V. P. (2011). *Designing and conducting mixed methods research.* Sage Publications.
Cukier, K., & Mayer-Schoenberger, V. (2014). *The rise of big data: How it's changing the way we think about the world.* Princeton University Press.
Davis, E. (2015). *TechGnosis: Myth, magic, and mysticism in the age of information.* North Atlantic Books.
Galimberti, U. (2023). *L'etica del viandante.* Feltrinelli Editore.
Giacomin, J. (2014). What is human centered design? *The Design Journal, 17*(4), 606–623.
Holmes, K. (2020). *Mismatch: How inclusion shapes design.* MIT Press.
Khanna, A., & Khanna, P. (2013). *Hybrid reality: Thriving in the emerging human-technology civilization.* TED Books.
King, R., Churchill, E. F., & Tan, C. (2017). *Designing with data: Improving the user experience with a/B testing.* O'Reilly Media.
Lovelock, J. (2000). *The ages of Gaia: A biography of our living earth.* Oxford University Press.
Lu, J., Gomez Ortega, A., Gonçalves, M., & Bourgeois, J. (2021). The impact of data on the role of designers and their process. In *Proceedings of the Design Society* (Vol. 1, pp. 3021–3030).
Nielsen, J. (2016). *The distribution of users' computer skills: Worse than you think.* Nielsen Norman Group.
Norman, D. A. (2023). *Design for a better world: Meaningful, sustainable, humanity centered.* MIT Press.
Nosta, J. (2017). *Technology's Battle for IQ, EQ, or something very different.* Psychology Today.
OECD, Organisation for Economic Co-operation and Development. (2016). *Skills matter: Further results from the survey of adult skills, OECD skills studies.* OECD Publishing.
Ogonowski, C., Ley, B., Hess, J., Wan, L., & Wulf, V. (2013). Designing for the living room: Long-term user involvement in a living lab. In *Proceedings of the SIGCHI conference on human factors in computing systems* (pp. 1539–1548).
Perez, C. C. (2019). *Invisible women: Data bias in a world designed for men.* Abrams.
Pheasant, S., & Haslegrave, C. M. (2005). *Bodyspace: Anthropometry, ergonomics and the design of work.* CRC press.
Rek, M., Romero, N., & Boeijen, A. V. (2013). Motivation to self-report: Capturing user experiences in field studies. In *Latin American conference on human computer interaction* (pp. 111–114). Springer.
Rohrer, C. (2014). *When to use which user-experience research methods.* Nielsen Norman Group.
Romero Herrera, N. (2017). The emergence of living lab methods. In *Living labs* (pp. 9–22). Springer.
Rubin, J., & Chisnell, D. (2008). *Handbook of usability testing: How to plan, design and conduct effective tests.* John Wiley & Sons.
Sanders, E. (2005). Information, inspiration and co-creation. In *Proceedings of the 6th international conference of the European academy of Design University of the Arts.* Bremen.
Slavin, K. (2016). Design as participation. *Journal of Design and Science, 1*(1).
Spinuzzi, C. (2005). The methodology of participatory design. *Technical Communication, 52*(2), 163–174.
Tosi, F. (2018). *Ergonomia and Design. Design per l'Ergonomia.* FrancoAngeli.
Vacanti, A. (2024). *Convivere con l'artificio: metodi e pratiche data-driven per il progetto dell'interazione.* Genova University Press.
Van Kleef, E., Van Trijp, H. C., & Luning, P. (2005). Consumer research in the early stages of new product development: A critical review of methods and techniques. *Food Quality and Preference, 16*(3), 181–201.
Van Mensvoort, K. (2022). *NextNature. Perché la tecnologia è la nostra natura del futuro.* D Editore.
Veeckman, C., & Van Der Graaf, S. (2015). The city as living laboratory: Empowering citizens with the citadel toolkit. *Technology Innovation Management Review, 5*(3).
Wang, T. (2016). *Why big data needs thick data.* Ethnography Matters.
Westerlund, B. (2007). A workshop method that involves users talking, doing and making. In *Proceedings of international conference on human-machine interaction, Human07, IEEE.*

Adaptable Housing in Slovak Practice

Lea Rollová, Zuzana Čerešňová, and Andrea Bacová

Abstract

The paper deals with research on adaptable housing for the needs of Slovak practice, which is a necessary response to negative social trends, especially the aging of the population. Research focusing on adaptable housing was carried out within two national projects in 2019–2024 and was divided into three phases: (Aday et al., 2018) examining the application of adaptable housing in European practice and legislation, (Larsson Ranada & Österholm, 2022) examining the principles of adaptable housing using the "research by design" method and participatory planning, the output being 9 model architectural studies of adaptable housing, a methodology Universal design of social care facilities, and checklists to evaluate residential building projects. The results of the second phase of the research were used as annexes to the call for submission of investment projects implemented from the Recovery and Resilience Plan. In the final (Maisel et al., 2008) phase, the investment projects submitted for buildings suitable for providing care services were evaluated. Of the 59 evaluated projects, more than half met the required evaluation criteria. The results of the presented research were used as a basis for subsequent adjustments to the already mentioned methodology and checklists, as well as in the creation of evaluation criteria within the framework of the evaluation of investment projects implemented in Slovakia from European sources.

Keywords

Universal design · Adaptable housing · Social care

1 Introduction

Slovakia is one of the fastest aging populations in Europe, and therefore it is very important to address issues related to sustainable housing that can better meet the needs of older adults dependent on social care, as well as the requirements for long-term sustainability of the Slovak social system. The changes being gradually introduced in Slovakia are mainly related to the implementation of the UN Convention on the Rights of Persons with Disabilities and the current reform of the social care system. These include, for example, changes in the assessment of the degree of dependence on social care and changes in the system of financing social services. Greater emphasis is placed on the quality of person-

L. Rollová (✉) · Z. Čerešňová
Slovak University of Technology in Bratislava, Centre of Design for All—CEDA, Bratislava, Slovakia
e-mail: lea.rollova@stuba.sk

A. Bacová
Slovak University of Technology in Bratislava, Faculty of Architecture and Design, Bratislava, Slovakia

S. Camplone and C. Porfirione (eds.), *Design for All and the Future of the Planet*, Design For Inclusion 6, https://doi.org/10.1007/978-3-032-21769-1_6

centered services, which prefer the provision of services in the home environment according to the concept of Aging in Place, when it is necessary to create a network of community social services in the place of residence, especially supported housing and day centers (Aday et al., 2018; Larsson Ranada & Österholm, 2022). A prerequisite for the operation of the concept of aging in place is the accessibility of the built environment of public spaces and public buildings according to the principles of universal design. As a result of inaccessible buildings and difficulties in moving in public spaces between buildings, people with disabilities suffer from many disadvantages and barriers that hinder their inclusion (Rollová et al., 2023; Williams & Willmott, 2012).

The Centre of Design for All—CEDA at the Faculty of Architecture and Design STU in Bratislava was involved in two national projects in 2019—2024, which were cofinanced by The European Social Fund and the Ministry of Labor, Social Affairs and Family of the Slovak Republic (hereafter referred to as ministry of social affairs). The first national project 'Deinstitutionalization of Social Service Facilities—Support for Transformation Teams' (hereinafter referred to as DI STT) focused on the transformation of social services in Slovakia, which consists of creating a wide range of new community social services, including various types of housing for recipients of social services, especially supported housing (MLSAF SR, 2016). This national project was carried out in cooperation with social workers experts. The research of the second national project Support for Universal Design (hereinafter referred to as SUD) focused on creating a proposal for systemic measures, with the aim of implementing the principles of universal design into national legislation and construction practice. The project was broadly conceived, with one of the five research teams focusing on research related to adaptable housing. Both projects were linked, particularly in the area of sustainable housing development, as several outputs of the DI STT project provided a suitable basis for research within the SUD project.

2 Methodology

The research focused on adaptable housing was divided into three phases: (Aday et al., 2018) examining the application of adaptable housing in European legislation and practice, (Larsson Ranada & Österholm, 2022) analyzing the principles of adaptability of residential buildings using the "research by design" method and participatory planning and (Maisel et al., 2008) analyzing and evaluating investment projects of buildings suitable for providing care services.

As part of the SUD project, **the first phase** of research was carried out, which focused on exploring technical standards and methodologies for designing adaptable housing, as well as positive examples from practice that have been published in European countries. The methods used were mainly a survey of available information on the web, selection and visit of selected buildings, and interviews with architects and researchers in selected European countries. The result of this research phase was an analysis of the state of adaptable housing in European legislation and practice, as well as a selection of inspiring models and practices. For the purposes of further research, models of adaptable housing in Sweden, Finland, Austria, and Belgium were selected.

The research in **the second phase** focused on designing and verifying adaptable housing projects using the methods of "research by design" and participatory planning. In this phase, "in situ" activities were carried out within the DI STT project, particularly an analysis of the physical environment quality in 92 participating social service facilities using the verified WHO Quality Rights Tool Kit (WHO, 2019). The first topic "Right to an adequate standard of living" was relevant for the assessment of the physical environment. During the design research, interviews were conducted with care recipients and social service providers. The following results of this phase were used for the call for submission of project proposals to expand community care capacities, which are published on the website of the Ministry of social affairs (MLSAF SR, 2022a).

- 9 architectural studies of 'Housing for small groups of social care recipients' (maximum 12 recipients), in which the principles of adaptable housing were applied.
- Methodology 'Universal design of social care facilities', including checklists.

The final phase aimed to verify the accuracy of the outputs from the second phase. The verification took place in the form of an analysis of 59 investment projects that responded to the call from the ministry of social affairs for the submission of project plans. When analyzing the project documentation of the buildings, it was examined whether the standards of universal design and adaptable housing were met. Finally, the most common reasons for negative evaluation of investment projects were processed and comments from the professional community were incorporated, which were the basis for subsequent adjustments to the methodology Universal design of social care facilities and checklists.

3 Research Findings

In **the first phase**, the research focused on the implementation of the concepts of adaptable housing based on anticipated changes during life and creation of adequate conditions for caring for older people or a family member with a disability. Legislation and standards in the field of accessibility focused on adaptable housing in several European countries were analyzed. Based on the data collected, the concept of adaptable housing is being implemented in several countries, especially in the Nordic countries in Europe. Some countries have defined requirements for adaptable housing in legislation and standards, e.g. Sweden, Finland, Belgium, Austria. Adaptable housing is also covered in Chapter 15.4. EN 17210:2021 Accessibility and usability of the built environment—Functional requirements.

For example, in Sweden, all new construction of apartment buildings and the surrounding environment since 2010 has been implemented in accordance with the concept of adaptable housing. The residential environment is designed to allow not only accessibility and adaptability of the apartment, but also active participation in various activities in the community, such as barbecues with neighbours, physical activities on multi-generational playgrounds, walks and relaxation with friends. Swedish regulations and standards served as the basis for the preparation of documents for the creation of Slovak regulation for accessible built environment, especially for the requirements for adaptable bathrooms. According to Swedish regulations, all new and renovated apartment buildings must be accessible, and each apartment unit must be adaptable in a way that does not require additional major construction modifications, but only minor adjustments. This considers not only the possibility of remaining in one's home environment in case of a change in health status, but also the principle of "visitability". This is very important factor in creating adaptable living environments, according to Maisel et al. (2008), the requirement of "visitability" aims to improve the ability of older people and people with disabilities to visit the homes of their family and friends. Such an environment does not create barriers for people of different ages, which significantly helps prevent isolation of older people and leads to social inclusion.

The Swedish standard SS 91 42 21 (Building design—Housing—Internal dimensions) specifies requirements for adaptable housing, including the dimensions of rooms and the placement of interior furnishings and equipment to meet the requirements for accessibility and usability. The aim is to improve conditions for older people and people with disabilities to remain at home and to provide care in the home. The dimensions of the spaces are based on the need for the smallest manoeuvring space for a person in a wheelchair, which according to Swedish regulations is a circle with a diameter of 1300 mm. Such manoeuvring space must be in the entrance areas of the apartment, at the kitchen counter, in the bathroom (together with the toilet), in at least one bedroom, in the living room and on the balcony. All windowsills are standardly at a reduced height (maximum 700 mm), which allows

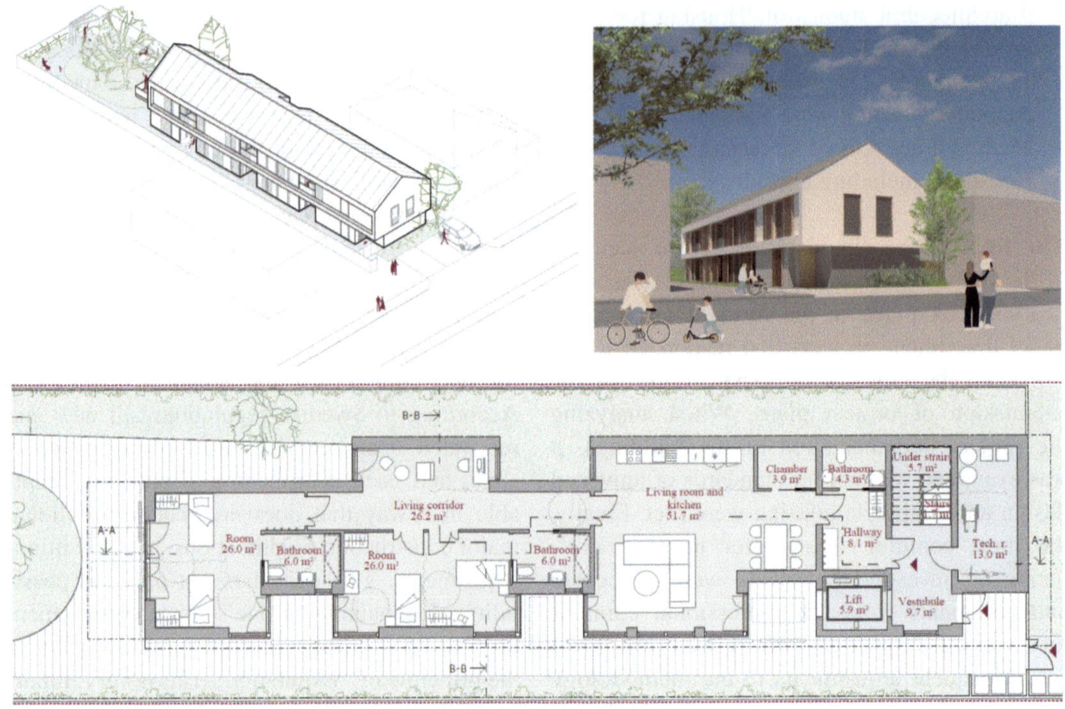

Fig. 1 Model architectural study TYP B of housing for small groups of social care recipients. Authors: Lea Rollová and Natalia Bošková Filová

a view even for a person sitting in a wheelchair or a person lying in bed. All doors, including balcony doors, must allow access for people in wheelchairs, therefore all doors have a clear width of at least 800 mm, and in the case of an adaptable bathroom, they open outwards from the room. At least one bathroom in the housing unit must be adaptable and usable for people with limited mobility and must be designed so that the space can be easily used even with a caregiver. Adaptable bathrooms have a floor with a drain and a shower area at floor level, with the toilet bowl located near the shower so that, if necessary, this space can be used to transfer a person in a wheelchair from the side of the toilet bowl (after adding the necessary handles). Such a solution provides an efficient use of space and is very easily adjustable, even in terms of the requirements of people in wheelchairs. The concept of adaptable housing allows the resident to remain in the apartment even if their health condition changes due to the natural aging process or a sudden illness or injury.

The second research phase focused on the practical application of the knowledge gained from European legislation and practice with the aim of designing suitable models of adaptable housing using the research-by-design method and participatory planning. Within the DI STT project, the various needs of social service recipients were investigated during interviews and surveys. The following findings were used when designing 9 model architectural studies of housing for small groups of social care recipients (see Fig. 1).

- Residential buildings for social care should have the appearance of standard housing in ordinary environment.
- Each care recipient has the right to privacy in his room; recipient can choose whether to live in the room alone or in a pair with a partner, friend, etc.
- The spatial design of apartments must also allow for the transport of lying recipients,

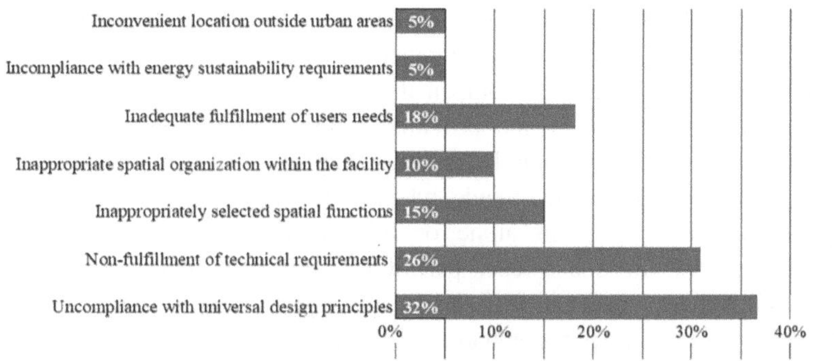

Fig. 2 The most common shortcomings of the analyzed investment projects. *Source* Rollová et al., 2024

especially to the social part of the apartment and to the exterior.

- There are several safety and security features that are essential in buildings, e.g. various intelligent technologies, and assistive technologies.
- Caregivers who provide care respect the rights to privacy of recipients and, therefore, cannot have their own space located directly in the recipients' apartment.

A selected model architectural study of housing for small groups of social care recipients was also published in the proceedings of the WMCAUS 2022 conference (Rollová & Bošková Filová, 2023). Key outcomes are also the methodology Universal design of social care facilities, and checklists to evaluate residential building projects. Adaptable apartments should already be prepared in the design phase to allow easy changes in a short time and at a low cost according to current needs. The evaluation criteria developed in the form of checklists were evaluated as an effective tool for self-monitoring by designers. The building design phase is of fundamental importance, as it is at this stage that most of the decisions are made that ultimately affect the general accessibility of the building and its surroundings, as well as the adaptability of apartments to the needs of apartment users.

In **the final phase** of the research, it was analyzed whether the outputs from the previous phases of the research were processed correctly, whether they are sufficiently understandable for designers, whether they can provide sufficient guidance on the correct application of mandatory elements, etc. Research in this phase focused mainly on the analysis of 59 investment projects that were submitted according to the call of the Ministry of Social Affairs (MLSAF SR, 2022b). When analyzing and evaluating investment projects, 30 of 59 fully met the call conditions and were therefore recommended for implementation, 14 out of 59 had to complete the missing information and only then were recommended for implementation, and 15 out of 59 were rejected due to failure to meet the conditions, which is as much as 25.41%. The reason for the rejection was fundamental errors in the application of the specified requirements, with the most shortcomings related to the application of universal design (see Fig. 2).

4 Discussion

The evaluation of investment projects for the provision of community social services, which were implemented in Slovakia from the Recovery and Resilience Plan, brought several stimuli for reflection and for changing the measures established from the previous phases of research. For illustration, we present the experience from the implementation of the model architectural study TYPE I in practice, which was approved in 2023. This new social services facility in Okoč can be considered a model pilot project that demonstrates the viability and effectiveness of set procedures. Smaller facilities allow for better communication and an individualized approach to clients. In Okoč, two adaptable houses were built on a plot of land. Each house has rooms for six clients, two large bathrooms, a shared kitchen with a dining room, and a living room. In the opinion of social care recipients and providers, adaptability is a benefit, especially for care recipients, because they could decide whether they want to live in a room alone or in a pair with a friend. In one house, partitions were inserted between some rooms and in the other house three double rooms remained. The accessible garden and the spacious terraces with pergolas, which create a very pleasant living space and a contact zone for clients, were also positively evaluated. The pergolas conveniently connect both objects and divide the exterior space into several zones with the possibility of different uses, according to the needs and interests of the clients. This solution was also praised by care providers because they could move between houses without problems even in adverse weather. However, the fundamental shortcoming was that the architectural study was intended as supported housing, but in reality, two bedridden recipients also live in one of the houses, for whom the 900 mm wide door is not wide enough. Therefore, transporting bedridden recipients to the common room or to the garden is more complicated. All findings (obtained during the evaluation of 59 projects) were then reviewed and consulted with social workers. For example, when updating the conditions of the call, it was determined what types of social services must be designed to allow the transfer of bedridden care recipients, if necessary (this was especially the case in specialized facilities where care for bedridden recipients is expected). Here it is necessary to design a door width of 1100 mm. Changes have also been made to the methodology and checklists when designing the bathroom so that showering is possible in a lying position, if necessary.

5 Conclusion

Based on demand, it can be expected that in the future, adaptable housing will be applied in Slovakia to a much greater extent than so far. The results of the presented research were subsequently applied in the creation of national construction legislation, in particular the Construction Act and the Decree on accessible design and use of buildings, where the comment procedure is currently underway. As part of the discussion with construction experts, the obligation to build adaptable apartments in all new residential buildings was waived but the draft decree enshrines the obligation to build adaptable apartments if they are built with public funds. However, it is necessary to adopt minimum standards for all apartments so that people can be taken care of in them, if necessary. The new Slovak construction legislation is to be approved by the parliament in April 2025. Local policy instruments such as local building regulations should also ensure that people with disabilities and older people with various health disadvantages can remain in their homes according to the concept of aging in place.

References

Aday, R. H., Wallace, B., & Krabill, J. J. (2018). Linkages between the senior center as a public place and successful aging. *Activities, Adaptation & Aging, 43*(3), 211–231. https://doi.org/10.1080/01924788.2018.1507584

Larsson Ranada, Å., & Österholm, J. (2022). Promoting active and healthy ageing at day centers for older people. *Activities, Adaptation & Aging, 46*(3), 236–250. https://doi.org/10.1080/01924788.2022.2028050

Maisel, J. L., Smith, E., & Steinfeld, E. (2008). *Increasing home access: Designing for Visitability*. AARP, 2008.

MLSAF SR (2016) National Action Plan for transition from institutional to community-based care in the social services system 2016–2020. Bratislava, The Ministry of Labour, Social Affairs and Family of the

Slovak Republic. https://www.employment.gov.sk/files/slovensky/rodina-socialna-pomoc/socialne-sluzby/national-action-plan-transition-from-institutional-community-based-care-social-services-system-2016-2020.pdf

MLSAF SR (2022a) Catalogue of architectural studies. The Ministry of Labour, Social Affairs and Family of the Slovak Republic. https://www.employment.gov.sk/sk/uvodna-stranka/plan-obnovy-odolnosti/vyzva-predkladanie-projektovych-zamerov/katalog-architektonickych-studii/

MLSAF SR (2022b) Call for submission of project proposals from the recovery and resilience plan. The Ministry of Labour, Social Affairs and Family of the Slovak Republic. https://www.employment.gov.sk/sk/uvodna-stranka/plan-obnovy-odolnosti/vyzva-predkladanie-projektovych-zamerov/

Rollová, L., & Bošková Filová, N. (2023). Adaptable housing in the context of universal design and social care. *AIP Conference Proceedings, 2928*(1), 200002. https://doi.org/10.1063/5.0170565

Rollová, L., Hubinský, P., & Bošková Filová, N. (2023). Universal design and social care: Assistive robots as other users of the built environment? *Arch Pap Fac Arch Des STU, 28*(3), 10–17. https://doi.org/10.2478/alfa-2023-0015

Rollová, L., Končeková, D., & Tóthová, Z. (2024). Evaluation of a continuing education programme on socially responsible architecture. *Glob J Eng Education, 26*(3), 167–173.

Swedish Standard SS 91 42 21 (n.d.) *Byggnadsutformning—Bostäder—Invändiga mått* [Building design—Housing—Interior dimensions]. https://www.sis.se/produkter/byggnadsmaterial-och-byggnader/byggnader/bostadshus/ss9142212006/

WHO. (2019). *WHO quality rights materials for training, guidance and transformation.* World Health Organization, 2019. https://www.who.int/publications/i/item/who-qualityrights-guidance-and-training-tools

Williams, G., & Willmott, C. (2012). Higher levels of mobility are associated with greater societal participation and better quality-of-life. *Brain Injury, 26*(9), 1065–1071. https://doi.org/10.3109/02699052.2012.667586

Mapping Design for All (DfA) Through Ecological Frameworks to Co-design Inclusive Ecologies

Janice Rieger

Abstract

This paper presents global case study research on how through ecological thinking we can rethink and transform Design for All (DfA). By investigating DfA through a more holistic ecological framework this world first study will create new knowledge, practices, and theories for DfA renewal. Mapping DfA through an inclusive ecology means examining both the micro-level and macro-level ecologies that influence how we design for all. This work presents qualitative research in museums, galleries, exhibitions, and cultural heritage sites (in Canada, UK and Australia) to evidence how co-design coupled with ecological and systems thinking can shift our ways of knowing and doing DfA. These global studies mapped DfA through an ecological framework through national surveys, interviews, field work and observations, to create future trajectories for inclusive design and DfA. In design and architectural practice and research, we tend to focus on the product or process, or on the technology or user experience, but rarely do we map the entire ecosystem of DfA. This research does just that, by presenting new models that can shift our ways of doing and thinking in DfA in a more than human world, and how to map these entangled connections across typlogies. Redefining DfA through a more holistic ecological approach has the potential to make a significant and timely contribution to creating inclusive ecologies across diverse typologies like museums and galleries, office environments, urban spaces, schools, and health facilities to co-create socio-spatial justice for all.

Keywords

Co-design · Design for all · Ecological frameworks · Inclusive design · More than human

1 Introduction

Inclusive design and DfA are often created through codes, audits, guidelines or visitor surveys, and considered as an add on. This additive approach to design can only create surface level changes and not an inclusive ethos—an inclusive ecology.

Drawing upon ten years of research across Canadian, UK, European and Australian case studies and my research in the evolution of access and inclusion across typologies (e.g., art

J. Rieger (✉)
Associate Dean Research (Faculty of Design and Society) and Australian Research Council (ARC) Research Fellow, University of Technology Sydney, Sydney, Australia
e-mail: janice.rieger@uts.edu.au

S. Camplone and C. Porfirione (eds.), *Design for All and the Future of the Planet*, Design For Inclusion 6, https://doi.org/10.1007/978-3-032-21769-1_7

galleries, healthcare, museums, housing, urban design and planning), I have created best practice examples of co-designing inclusive ecologies beyond codes, manuals and guidelines alone, through an ecological approach for all.

2 Research Context and Challenges

This paper presents an ecological model for design where we can rethink Design for All (DfA) through a more holistic ecological framework by adapting Bronfenbrenner's Ecological Systems Theory (see Fig. 1). This world first study will create new knowledge, practices, and theories through reframing DfA through ecological thinking and doing. Mapping DfA through an inclusive ecology means examining both the micro-level and macro-level ecologies that influence how we design for all.

Creating inclusive ecologies is one that inscribes care in the materiality of more than human things. Like de la Bellacasa (2017), I too approach care as a "critically disruptive doing that can open to 'as well as possible' reconfigurations engaged with troubled presents." (pg. 12).

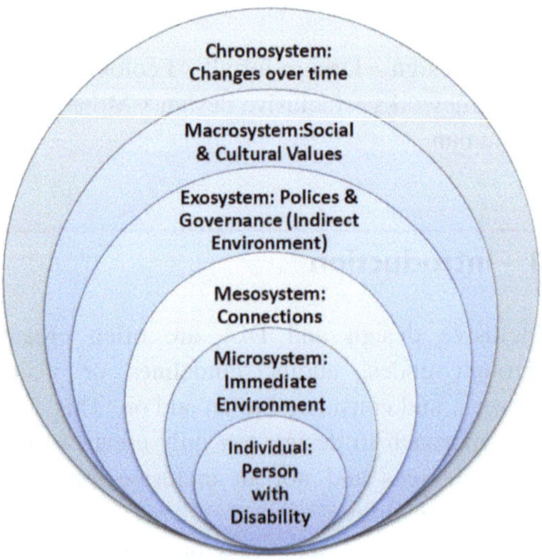

Fig. 1 Co-designing inclusive ecologies: adapted ecosystem model from Bronfenbrenner's ecological systems theory (Bronfenbrenner, 1979)

> Care is a human trouble, but this does not make of care a human-only matter. Affirming the absurdity of disentangling human and nonhuman relations of care and the ethicalities involved requires decentering human agencies, as well as remaining close to the predicaments and inheritances of situated human doings." (de la Bellacasa, 2017, pg. 2).

To address this gap in knowledge about the design, development, and implementation of DfA, this research aims to examine this significant issue more holistically, serving to advance an understanding of equal access by mapping an 'inclusive ecology' at the *micro level*, in daily actions and interactions, and at the *macro level*, in policies, systems, institutions, and infrastructure.

3 Process and Methods

This paper presents on a longitudinal research project from 2014-present, across Canada, USA, Europe and Australia, to evidence how ecological and systems thinking can shift our ways of knowing and doing DfA. Global case studies from 2014 mapped DfA through an ecological framework through national surveys, interviews, field work and observations, to create future trajectories for inclusive design and DfA.

Due to the nature of this paper, the full data and findings cannot be reported on but what I wanted to share in this paper was the findings that impact practice and to share examples of best practice (design, architecture, planning, etc.).

4 Over a Decade of Inclusive Ecologies for DfA through Global Case Studies

I first started to notice what an inclusive ecology could be while doing case study research at the Canadian Museum of Human Rights (CMHR) in Winnipeg, Canada in 2014. At this museum almost everyone in the instution from the janitorial staff, to the security guards to the accountants understood the importance of accessibility and inclusion, and were knowledgeable and educated around universal design principles, etc. (Rieger, 2016; Rieger, 2023). Unlike most institutions that have a sole inclusive champion, or an access and inclusion or DEI team, the CMHR shared this practice and knowledge and mandated that someone from every area of the musum be on the access and inclusion group (Rieger, 2023). The enthusiasm for inclusion and access was apparent throughout the entire institution and was effectively translated into all of the museums exhibitions, designs, services, operations, programs, spaces, facilites, training and archives—creating an inclusive ecology (Rieger, 2016; Rieger, 2023).

Five years later in Australia, in 2019 at the QUT Art Museum, I also witnessed first-hand and co-developed the creation of an inclusive ecology, and instead of designing access for museum visitors, QUT Art Museum co-designed access *with* them for DfA. The creation of an inclusive ecology impacted the museum's ethos, infrastructure, artists' creative practices, exhibition designs, wayfinding, installations, created employment opportunities for people with disabilities, and influenced other galleries across Australia. QUT Art Museum, like the CMHR in Canada five years earlier demonstrated how DfA can create a societal shift—an inclusive ecology (Hadley & Rieger, 2021; Rieger, 2020; Rieger et al., 2019; Rieger et al., 2022).

Fast forward another 5 years and in 2024, I am now conducting national and international research through an Australian Research Council Research Project (2023–2026) called *Co-creating Cultures of Inclusion: Redefining Access to Cultural Heritage.* As a world-first study this project will create an innovative co-design model of practice, through an ecological framework and inclusive multi-sensorial explorations that can be translated and adopted by national, state, university, and regional museums and galleries across Australia and globally. It was during my field studies in London that I again found the power of how an inclusive ecology can impact an entire institution (both human and nonhuman) to create DfA. This was at the Wellcome Collection in London, England, where every aspect of their museum from the micro to the macro, is about creating an inclusive ethos and an inclusive ecology.

5 An Inclusive Ecological Model for DfA Practice

These studies highlight the possibility of 'an inclusive ecology' to exist in a cultural institution. Wellcome Collection (UK), QUT Art Museum (Australia), and the Canadian Museum for Human Rights (Canada) exemplify in many respects an inclusive institutional ethos. However, what the vast majority of institutions have not embraced is a DfA ethos, in which an institution can be used and accessed by all, irrespective of their physical, sensory or intellectual abilities (Eardley et al., 2016; Rieger, 2020; Rieger, 2023). This was echoed through semi-structured interviews with museum staff in both Canada and Australia:

> Without a statement that is shared, understood and applied from the highest levels of the institution, creating inclusive experiences will remain within the domain of conforming to code, and the capacity, interest or commitment of individuals or teams (Senior Executive, Canadian Museum, 2015).
>
> What I have seen as the single biggest barrier to even meeting or surpassing standards, guidelines and best practices has been the absence of a kind of philosophical statement about the importance of inclusion (Senior Executive of National Institution in Australia, 2017).

Building upon this global evidence of the importance of a more holistic inclusive ecology that looks at both the human and nonhuman, I co-designed (with a national advisory group) a National Survey and disseminated the online survey in 2023 to practitioners across Australia. This Survey was delivered ~8 years after these aforementioned interview testimonials, but the findings are similar:

> Staff onsite very much want to improve accessibility, but the 'company' that runs the museum lists

> access issues as 'not a priority'. (National Survey Participant, Australia, 2023).
>
> (...) willful ignorance of leadership. (National Survey Participant, Australia, 2023).

The evidence from ten years of global case studies are that creating an inclusive ecology for/with/through DfA contrasted with the data and findings from institutions that list DfA and access as 'not a priority', and/or institutions that have access champions that are burnt out and not listened to, point to issues around DfA in practice that are enduring, and have been for decades.

6 Discussion

So, what does this mean for practitioners and for DfA? Do designers need to rethink their roles and responsibilities in creating DfA that goes beyond design? How can practice adapt to address current and emerging challenges and future opportunities?

Drawing upon *Gaia,* this paper presents a more holistic approach for design practice where we encourage care for all, not just Design for All—care for the earth, its inhabitants and all of life (human and nonhuman) in its complexity. Caring is a complex matter, and it is important to understand that caring is not just a *caring for*, perpetuating the issues with the charity model of disability but care is entangled with human and nonhuman things (Rieger, 2023). A new approach for DfA is needed to amplify the often silenced or forgotten voices of our users (and the nonhuman voices) to generate new models of practice for inclusion. We can create care-full design for all, where a reimagining of a duty of care is/becomes more than just care for people or a care for a building or a design or an artefact but a more holistic ecology of care, that can invoke embodied, and embedded relations with all things (Rieger, 2020; Rieger, 2023).

> Thus conceived, the complexity of the circulation of care feels even more all- pervasive when we think of how it is sustained in more than human worlds. Care is a force distributed across a multiplicity of agencies and materials and supports our worlds as a thick mesh of relational obligation. (de la Bellacasa, 2017, pg. 20).

This paper argues for a more holistic practice of design which includes a post humanist perspective or more than human approach to better understand inclusion beyond just design. Yes, I am arguing that DfA needs to move beyond a design and designer-centric approach. This paper advocates for the expansion of inclusive design into a more material, care-full, more than human approach to acknowledge the importance of diverse embodied knowledge and co-designing practices to create inclusive ecologies. In other words, by looking more holistically at DfA as an ecology with human and nonhuman actors and by mapping this ecology, from the micro to the macro (with our clients and users), a shifting of perspective can occur from something that is quite narrow in focus to something that expands (Rieger, 2023).

Failure to address long-term engagement, and the life of ecologies, including the sustainability of relations with all things, and the constant evaluation and an iterative approach will most likely result in tokenistic outcomes (Rieger, 2023).

In design practice and research, we tend to focus on the product or process, or on the technology or user experience, but rarely do we map the entire system or ecology of DfA. By redefining access through inclusive ecologies a curious thing happens— it creates an ethos of care and inclusion for human and nonhuman things alike. This shift not only in practice but attitude, and instutitional ethos, policies, procedures, infastructure, etc. (the macro-level of the ecology) also has the ability to scaffold and become sustainable for all (Rieger, 2023).

7 Conclusion

This paper explores the tensions and opportunities for DfA in a more than human world, and how to map these entangled connections across typologies to collaborate on new models of practice through an ecological framework.

The key takeaways from this chapter offer useful prompts to designers, architects, planners, curators, managers, educators and scholars to rethink how they are understanding DfA and co-designing more fully through inclusive ecologies. This has the ability to shift the knowledge production and practice of inclusive design and DfA towards an ecology of care for a more than human world.

References

Bronfenbrenner, U. (1979). *The ecology of human development: Experiments by nature and design*. Harvard University Press.

de la Bellacasa, M. P. (2017). *Matters of care: Speculative ethics in more than human worlds* (Vol. 41). University of Minnesota Press.

Eardley, A. F., Mineiro, C., Neves, J., & Ride, P. (2016). Redefining access: Embracing multimodality, memorability and shared experience in museums. *Curator: The Museum Journal, 59*(3), 263–286.

Hadley, B., & Rieger, J. (2021). Co-designing choice: Objectivity, aesthetics and agency in audio-description. *Museum Management and Curatorship, 36*(2), 189–203. https://www.tandfonline.com/doi/full/10.1080/09647775.2021.1878469

Rieger, J. (2023). *Design, disability and embodiment: Spatial justice and perspectives of power*. Taylor & Francis.

Rieger, J., Hadley, B., Barron, S., Boulton, S., & Parker, C. (2022). Codesigning access: A new approach to cultures of inclusion in museums and galleries. In *Curating access: Global Exhibitions and Creative Accommodation* (pp. 183–195). Routledge.

Rieger, J. (2016). Doing dis/ordered mapping/s: Embodying disability in the museum environment, University of Alberta. https://era.library.ualberta.ca/items/72964556-d167-4126-8373-f1a97931185b

Rieger, J., Van Ooyen, V., Dionysius, K., & Rafferty, L. (2019). Vis-ability: Artwork from the QUT art collection. https://eprints.qut.edu.au/132875/

Rieger, J. (2020). Moving beyond visitor and usability studies: Co-designing inclusion in museums and galleries. In *Advances in industrial design: Proceedings of the AHFE 2020 virtual conferences on Design for Inclusion, affective and pleasurable design, interdisciplinary practice in industrial design, Kansei engineering, and human factors for apparel and textile engineering, July 16–20, 2020, USA* (pp. 167–172). Springer International Publishing.

Academic Research: Products and Services

Open Your Mind. Mental Health, Not Mental Illness. Promoting Mental Health Through Design

Xavier Ferrari Tumay and Federica Maria Lorusso

Abstract

This paper examines the collaborative initiative 'Apri la mente' (Open your mind), a project developed by the Day Centre for Mental Health DSS 12 in Genoa and the Department of Architecture and Design of the University of Genoa. The project aims to reduce the stigma associated with mental health disorders through a participatory design approach, to foster patients' empowerment and promote their social and emotional integration. The initiative used workshop activities focusing on creative writing, 3D printing, screen printing and designing handicrafts to stimulate mental rehabilitation processes and strengthen the participants' autonomy. The project aligned with the principles of 'serious leisure', demonstrating the therapeutic potential of artistic expression in processing emotions and increasing self-awareness. It was structured in three phases: (1) creative writing workshops allowed participants to narrate their own mental health experiences, fostering mutual understanding and empathy; (2) hands-on workshops at the University of Genoa's Digital Lab emphasised 'learning by doing', enabling participants to acquire technical skills in 3D printing and screen-printing; (3) participants collaborated with university students to design an exhibition stand and sell their creations during a local fair, promoting social inclusion and economic independence. The initiative underlines the value of collaboration between sectors, creating a bridge between academia and mental health services to tackle stigma and support the rehabilitation of patients. The project also highlighted the importance of relational and emotional skills in psychiatric care, improving participants' self-esteem and sense of self-efficacy. The success of this project is confirmed by comparing it with similar initiatives such as MinD—Mad in Design and Patternhouse, which demonstrate how co-design can be an effective strategy to foster psychiatric rehabilitation and raise awareness of the importance of mental health, reducing social isolation and building sustainable community support systems.

Keywords

Psychiatric disorders · Co-design · Empowerment · University third mission · Communication

X. F. Tumay (✉)
Architecture and Design Department, University of Genoa, Genoa, Italy
e-mail: xavier.ferraritumay@unige.it

F. M. Lorusso
Architecture and Design Department, University of Campania Luigi Vanvitelli, Caserta, Italy
e-mail: federicamaria.lorurro@unicampania.it

S. Camplone and C. Porfirione (eds.), *Design for All and the Future of the Planet*, Design For Inclusion 6, https://doi.org/10.1007/978-3-032-21769-1_8

1 Introduction

This article explores the collaborative initiative "Apri la mente" (Open Your Mind), developed jointly by the Day Center for Mental Health DSS 12 in Genoa and the Department of Architecture and Design at the University of Genoa. The primary objective of this project is to mitigate the persistent stigma associated with mental health through the implementation of a co-design approach. The shared aim is to normalize the emotional states related to psychiatric disorders, emphasizing the importance of perceiving these conditions as manageable and treatable.

Participatory design, by its very nature, enables patients to feel integral to the therapeutic process, fostering a sense of autonomy and empowerment. Recent studies have demonstrated that the active involvement of patients in the design of therapeutic interventions can significantly enhance clinical outcomes. For instance, an analysis conducted by Simonsen and Robertson (2012) highlights how participatory design can improve patients' quality of life by facilitating the development of solutions better tailored to their specific needs. Additionally, research by Hamilton and Tetu-Mouradjian (2015) indicates that patients who actively participate in the design of their care pathways report higher levels of satisfaction and adherence to proposed treatments.

The openness of the university and its community to engage beyond academic boundaries underscores an additional level of significance. The collaboration with the university extends the initiative's impact beyond its immediate context and fosters a productive dialogue between academia and practice within the local setting. This dialogue aligns with prior collaborations initiated by DINOGMI (the Department of Neurosciences, Rehabilitation, Ophthalmology, Genetics, and Maternal and Infant Sciences) as part of the Mental Health Pact "The Caring City" (2017). Such synergy facilitates the exchange of knowledge and perspectives, contributing to mutual enrichment.

Including patients in university environments introduces a layer of personal and pragmatic relevance. This approach aims to address the specific challenges faced by individuals living with psychiatric disorders directly in academic environments, promoting awareness of this group's needs and providing concrete resources to support them.

2 Scenario

In Italy, prior to the 1980s and the process of deinstitutionalization, traditional approaches to mental health care primarily focused on symptom management and medical treatment, often neglecting the potential of individuals to regain functional abilities and reintegrate into their communities (Burti, 1994). In 1978, neurologist and psychiatrist Franco Basaglia pioneered psychiatric reform to advocate for the rights of individuals with mental illnesses, culminating in the enactment of Law 180. This healthcare legislation embraced a more humane approach compared to traditional asylums institutions, promoting the closure of psychiatric hospitals and the establishment of territorial mental health services designed to facilitate the integration of psychiatric patients into society.

In the United States, two decades before Italy's deinstitutionalization, John Beard, founder of Fountain House, emphasized that regardless of the severity of mental illness and its associated limitations, it was essential to focus on the functional capacities of each individual, as every patient retains a core of essential ego strengths (Beard et al., 1958). This perspective remains a cornerstone of Fountain House's philosophy, one of the leading American organizations providing support and rehabilitation for individuals with mental illnesses (Goertzel et al., 1960). Institutions like Fountain House aim to enhance patients' health and well-being through participatory activities, fostering the restoration of lost capacities, reducing social isolation, and increasing opportunities for connection with the external world.

In the context of mental health, healthcare facilities such as Community Mental Health Centers provide environments conducive to the recovery and self-awareness of patients regarding their unique identities and abilities. These centers offer individuals with psychiatric disorders a range of rehabilitative services aimed at improving or developing essential behavioral and relational skills, thereby facilitating their reintegration into society. Within these settings, healthcare professionals, including therapists, health workers, and caregivers, establish rehabilitation objectives through therapeutic and rehabilitative approaches that focus not only on treating the patients but also on actively caring for them. Beard et al. (1958) assert that therapeutic group activities that do not exceed the patient's capabilities can significantly complement traditional therapeutic efforts, as they allow the effective development of new social relationships, while simultaneously reducing symptomatology and pathology.

Indeed, rehabilitative activities achieve two primary objectives: the first is to enable patients to become autonomous in everyday tasks such as grocery shopping, interacting with others, or performing small jobs; the second is to promote the social inclusion of these individuals in groups outside the mental health day centers. However, rehabilitation activities are often limited to groups composed exclusively of patients, which reduces opportunities for external social inclusion due to limited contact with individuals outside the centers. To promote autonomy and social integration, it is essential for users of mental health day centers to participate in proactive activities such as support groups, occupational therapy sessions, and other initiatives.

A critical component of promoting mental health is the fight against prejudice, stigma, and discrimination often associated with mental illnesses. Mental disorders have historically been subject to greater stigma and negative judgment compared to other types of illnesses. Even today, widespread ignorance about nature and implications of mental disorders constitutes a significant societal issue, as a substantial portion of the population remains inadequately informed about these conditions and continues to perpetuate negative stereotypes and discriminatory behaviors against affected individuals. Consequently, many patients not only have to cope with the often devastating effects of their illness but also have to bear the brunt of social exclusion and prejudice (Rössler, 2016), further hindering their recovery and reintegration into society. Addressing this issue requires efforts to combat stigma and educate communities about various mental disorders through clear and accessible information, thereby fostering greater understanding and awareness of these conditions. Communities must mobilize to reduce social isolation and ensure that individuals with mental illnesses can lead connected and healthy lives, enabling them to achieve well-being.

According to the World Mental Health Report (2022), mental disorders are steadily increasing at both the local and national levels. The most prevalent and well-known conditions include anxiety, depression, bipolar disorder, schizophrenia, eating disorders, and borderline personality disorder.

In the specific context of the city of Genoa, data on mental disorders—kept confidential for privacy reasons—mirror both national and global trends. Notably, in recent years, there has been a decrease in the average age of patients, accompanied by an increase in admissions to mental health centers. One of the factors contributing to this rise is the forced isolation imposed during the COVID-19 pandemic, which significantly impacted the mental health of the population, particularly children and adolescents (Racine et al., 2020).

3 Co-Design and Mental Rehabilitation

Certain organizations employ participatory activities as an effective tool for the mental rehabilitation of individuals with mental health disorders. This approach not only helps to reduce the stigma associated with mental health but also enhances the well-being of these individuals. Through active involvement in such activities, participants receive support in their rehabilitation journey while also gaining greater self-confidence and trust in their own abilities and potential.

Two noteworthy Italian case studies highlighting the close connection between co-design and mental rehabilitation are *MinD—Mad in Design* and *PatternHouse.*

3.1 MinD—Mad in Design

MinD—Mad in Design is a cultural association established in Turin in 2019, stemming from an earlier concept developed by psychologist Elena Varini and architects Giulia Mezzalama and Sandra Poletto. The association addresses situations of distress and marginalization linked to mental fragility through the implementation of creative, multidisciplinary, and inclusive projects that use participatory design both as a transformative tool to improve the local context and as a means to promote social inclusion and rehabilitation through the creative process.

Currently, MinD acts as a cultural, educational, professional, and social inclusion space that directly engages vulnerable and fragile individuals supported by mental health services through a series of workshops and related events. In these events, the association fosters positive relationships aimed at empowering patients and recognizing their skills. As well as exploring an inclusive, multidisciplinary, and intergenerational approach involving the broader community.

By employing various methods of participatory and inclusive design, MinD serves not only as a valuable ally in the rehabilitative strategies for individuals with mental health disorders but also provides opportunities for redemption for marginalized people, by highlighting their inherent strengths and resources.

Each year, in collaboration with university students from across Italy, patients under mental health care, psychologists, social and healthcare workers, designers, architects, and local associations, MinD organizes workshops addressing mental health from multiple perspectives, with the aim of breaking down the barriers of prejudice and indifference towards fragile people.

Over the years, thanks to numerous workshops and the involvement of a wide range of stakeholders, MinD has been able to rethink the needs of psychiatric patients and the quality of mental health places, to investigate the role of cultural heritage for mental health, to experiment new solutions for domestic space and inclusive living, to enhance the relationships between people and places, places of care and places of culture, as well as redesign green areas in the municipality of Turin to make them inclusive for patients with mental disorders.

3.2 Patternhouse

Patternhouse is a research-based design project aimed at innovating the rehabilitation process for individuals with mental health conditions. Founded and creatively directed by Martina Drechsel, the project was launched in 2018 for the rehabilitation workshops of Casa Basaglia, a residential psychiatric rehabilitation facility managed by the Merano health authority in South Tyrol, Italy, with the desire to showcase the skills and talents of people with mental disorders.

Owing to its success, Patternhouse quickly evolved into a therapeutic method centered on creative, manual, and craft-based activities that patients could easily implement. The philosophy underpinning Patternhouse is rooted in the belief that manual and creative activities can reactivate individual capacities, revive personal skills, and foster mental rehabilitation and social interaction. A distinguishing feature of Patternhouse is its adaptability to the specific needs of each participant. By engaging in straightforward craft practices, the project seeks to create a stress-free environment where every individual, regardless of skill level or prior experience, can successfully participate in engaging and stimulating activities.

At the heart of the Patternhouse method is the creation of patterns, both on paper and fabric, using basic shapes and colors in an open-ended process that encourages exploration and creativity. During the initial stages of the creative process, participants design their own patterns by following three simple steps: copying, coloring, and recombining. This results in a personal album that serves both as a source of inspiration and as a catalog for selecting patterns to weave onto artifacts such as cushions. The project also includes the sale of these handcrafted products, with proceeds reinvested into the initiative to support its growth and further development.

4 Open Your Mind!

4.1 Project Objectives

The collaborative initiative "*Apri la mente*" (Open Your Mind), promoted by the Mental Health Day Center DSS 12 in Genoa and the Department of Architecture and Design at the University of Genoa, aimed to mitigate the persistent stigma associated with mental health through the implementation of several participative activities involving both mental health center patients and students of the University of Genoa. The activities undertaken focused on three key pillars: (i) empowerment of patients with mental health disabilities, fostering self-confidence and autonomy, (ii) co-design as an effective tool for facilitating mental rehabilitation, leveraging collaborative creative processes, (iii) third mission, extending the university's role beyond traditional academic boundaries to actively engage with the broader community.

4.2 Process of the Project

The project took place between November and December 2024 and was structured into three distinct phases.

The first phase involved a series of creative writing workshops, providing participants from the mental health center with an opportunity to share their personal stories and treatment journeys. Through this process, participants were able to define the concept of mental health in subjective and unique ways, culminating in the articulation of "representative phrases" that reflected the essence of their individual experiences.

These workshops not only offered a safe space for personal expression but also fostered a deeper understanding of mental health as a complex and multifaceted phenomenon. Sharing personal stories helped create an empathetic connection among participants, fostering a sense of community and mutual support.

The value of such activities is well-supported in the scientific literature. For instance, a study published in the *Journal of Clinical Psychology* highlighted the significant positive impact of creative expression activities, such as writing, on the emotional and psychological well-being of individuals facing mental health challenges (Pennebaker, 1997). Additionally, therapeutic storytelling, as discussed by White and Epston in their work on narrative therapy, allows individuals to reframe and restructure their life experiences, helping them find meaning and purpose in their personal narratives (White & Epston, 1990).

Creative writing serves as a medium to explore the motivations, characteristics, and impact of these activities on those who engage in them without the intention of becoming professional writers. These activities align with the concept of *serious leisure*—an activity pursued with dedication and commitment while not aimed at financial gain (Stebbins, 2006). These motivations include the desire for personal expression, the search for self-improvement, the need to tell their stories and the wish to communicate ideas and feelings.

Writing thus becomes a way to explore one's inner world and connect with others. Participants develop a critical approach to their own work and that of others, allowing them to appreciate the quality of the work and recognise any errors or room for improvement. Writing also offers a unique opportunity for introspection and personal growth, allowing individuals to explore their thoughts and emotions and achieve greater self-awareness.

From a psychological perspective, writing can have a therapeutic effect (Mazzara, 2003), promoting the processing of complex emotions

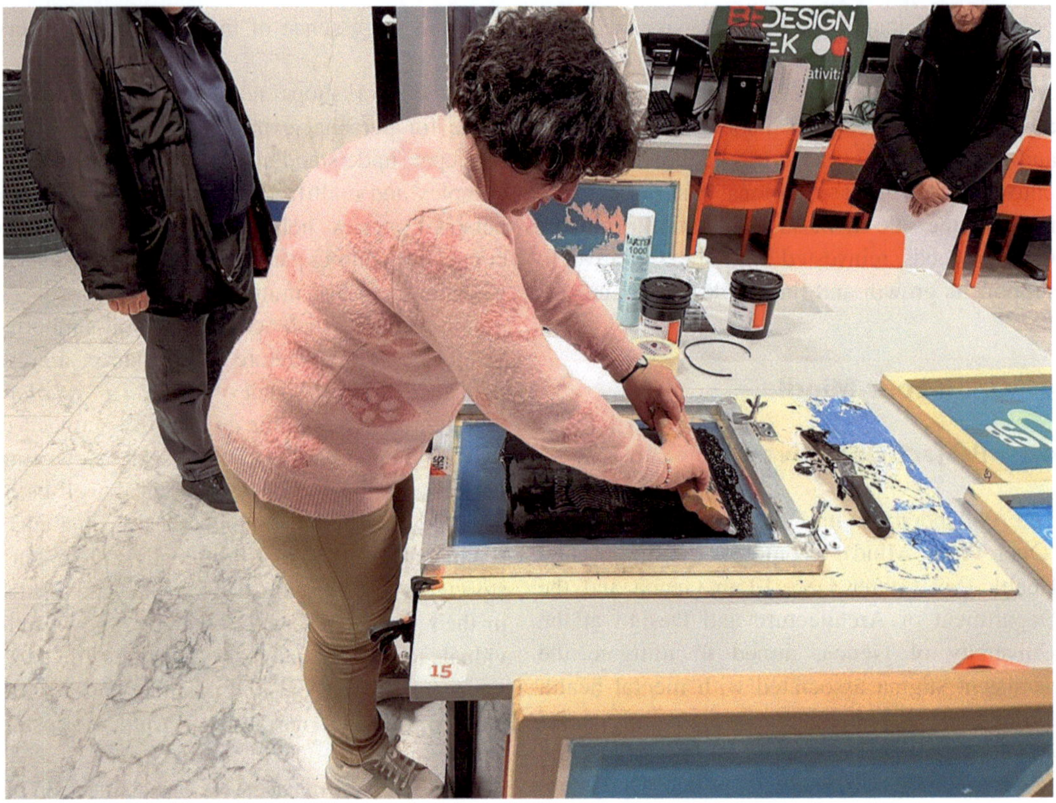

Fig. 1 Activities carried out during the workshop at the Digital Lab of the University of Genoa

and the overcoming of difficult periods. Furthermore, by sharing one's work within a community, it allows one to receive constructive feedback, by creating strong social bonds and finding support and encouragement.

The concept of *serious leisure* found practical application in the second phase of the project, held at the Digital Lab of the University of Genoa, where two workshops were conducted with mental health center participants. These workshops provided participants with the opportunity to learn and experiment with 3D printing and screen-printing techniques and production processes. These activities align perfectly with the principles of *serious leisure*, as they require commitment, the acquisition of new skills, and significant personal engagement (Fig. 1).

The "learning by doing" approach (Santos Arias, 2021) adopted in the workshops enabled participants to acquire knowledge and skills through direct experience, fostering active and participatory learning. This educational method, characteristic of Fab Lab communities—a model that the Digital Lab draws inspiration from—not only encourages self-directed learning and autonomy but also stimulates creativity and innovation. In this context, creative writing and artistic production serve as tools of expression for individuals with mental health disorders, providing them with an alternative means of communication.

Art, in all its forms, becomes a medium through which emotions and thoughts, which may be difficult to express through conventional communication channels, can be externalized. This artistic expression not only contributes to the psychological well-being of participants but also facilitates a deeper understanding and acceptance of their mental health conditions. In the third and final phase, design students from the Department collaborated to create an exhibition stand for participation in the *Mercati di S.*

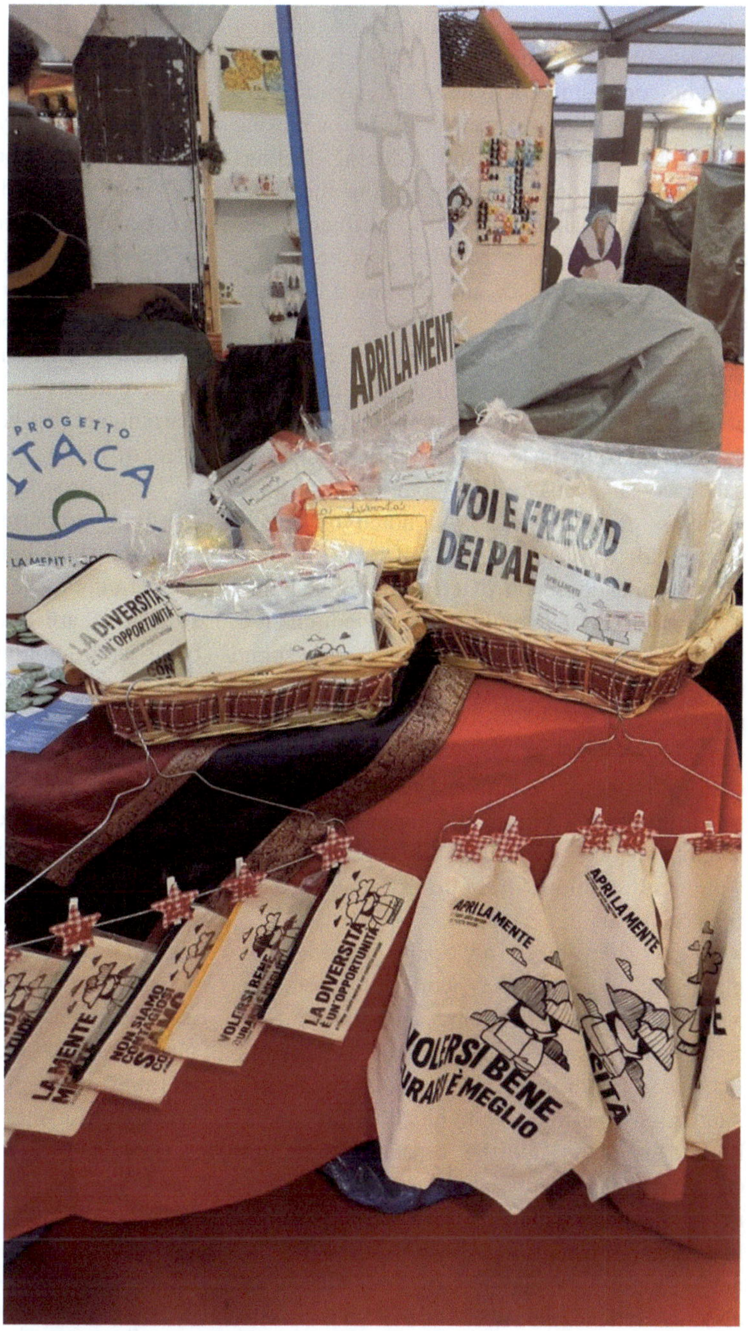

Fig. 2 Exhibition setup for the sale of products created during the activities

Nicola, a local event renowned for the sale of handcrafted products. During this phase, mental health center users had the opportunity to showcase and sell the material they had created during the creative writing workshops. This initiative served a dual purpose: to fund the Center's activities and to promote the social integration of its participants (Fig. 2).

Participation in community events such as the *Mercati di S. Nicola* offers mental health center users a meaningful platform to showcase their creative abilities, thereby enhancing their sense of self-esteem and belonging. According to a study by Harter et al. (2003), participation in community activities can significantly improve psychological well-being by increasing one's sense of identity and reducing feelings of social isolation. Moreover, selling the products created enables users to contribute financially to the center's activities, promoting a sense of self-sufficiency and economic independence.

The importance of self-financing for third-sector organizations is well-documented. The ability to generate income autonomously allows these organizations to maintain financial independence and ensure the long-term sustainability of their activities. As noted by Eikenberry and Kluver (2004), self-financing can reduce dependence on external funding, which is often subject to political and economic fluctuations, and allows organizations to respond more flexibly and innovatively to community needs.

The open book of social innovation (Murray et al., 2010) further emphasizes that economic self-sufficiency is crucial for social innovation, enabling organizations to experiment with new solutions without external financial constraints.

Events like these not only provide visibility for the creative work of users but also contribute to reducing the stigma associated with mental health disorders. As highlighted by Thornicroft et al. (2016), positive interactions between individuals with mental health disorders and the broader community can lead to greater social acceptance and a reduction in prejudice. Active participation in local markets and fairs allows users to be seen not only through the lens of their illness but also as individuals with talent and creative abilities. Finally, the involvement of design students from the Department in creating the exhibition stand fosters cross-sector collaboration, enriching both the students and the center's users. This synergy facilitates mutual learning and the development of social networks, which are essential for community support and growth.

5 Conclusion

The project demonstrated that inclusivity and intergenerationality are key elements in creating a dynamic and enriching learning environment. By involving individuals from different age groups and backgrounds, the project was able to draw on a wide range of experiences and perspectives, thereby enriching the co-design process. This approach not only led to more innovative solutions tailored to the specific needs of the patients but also facilitated a bidirectional learning process between university students and users of the Day Center for Mental Health.

The approach adopted in the project emphasized the importance of developing not only technical skills but also relational and emotional ones. In the context of psychiatric rehabilitation, soft skills are crucial for promoting the well-being and autonomy of patients. Competencies such as empathy, effective communication, and collaboration not only improve patients' quality of life but also enhance their active participation in the care process.

Another significant aspect of the project was the creation of an inclusive, training-oriented university workshop. This environment allowed participants to collaborate on real projects, applying the skills they had learned and fostering a sense of community and belonging. The training went beyond theoretical knowledge, incorporating practical experiences that made learning more meaningful and applicable to everyday life.

References

Beard, J. H., Goertzel, V., & Pearce, A. J. (1958). The effectiveness of activity group therapy with chronically regressed adult schizophrenics. *International Journal of Group Psychotherapy, 8*(2), 123–136. https://doi.org/10.1080/00207284.1958.11508807

Burti, L. (1994). Deinstitutionalization and application of the Italian psychiatric reform: Evolution of a revolution. *Epidemiologia e Psichiatria Sociale, 3*(2), 91–94. https://doi.org/10.1017/S1121189X00003523

Eikenberry, A. M., & Kluver, J. D. (2004). The marketization of the nonprofit sector: Civil society at risk? *Public Administration Review, 64*(2), 132–140.

Goertzel, V., Beard, J. H., & Pilnick, S. (1960). Fountain house foundation: Case study of an expatients club. *Journal of Social Issues*.

Hamilton, A., & Tetu-Mouradjian, L. (2015). Participatory design in health and Well-being: Analysis and impact. *Journal of Health Services Research & Policy*.

Harter, L. M., Japp, P. M., & Beck, C. S. (2003). *Narratives, health, and healing: Communication theory, research, and practice*. Routledge.

Mazzara, G. (2003). *Parole che nascono libere. La scrittura creativa per la crescita e la terapia* (Vol. 5).

Murray, R., Caulier-Grice, J., & Mulgan, G. (2010). *The open book of social innovation*. Nesta.

Pennebaker, J. W. (1997). Writing about emotional experiences as a therapeutic process. *Psychological Science, 8*(3), 162–166. https://doi.org/10.1111/j.1467-9280.1997.tb00403.x

Racine, N., Cooke, J. E., Eirich, R., Korczak, D. J., McArthur, B., & Madigan, S. (2020). Child and adolescent mental illness during COVID-19: A rapid review. *Psychiatry Research, 292*, 113307. https://doi.org/10.1016/j.psychres.2020.113307

Rössler, W. (2016). The stigma of mental disorders: A millennia-long history of social exclusion and prejudices. *EMBO Reports, 17*(9), 1250–1253. https://doi.org/10.15252/embr.201643041

Santos Arias, F. (2021). De la Bauhaus al Fab Lab. La revolución digital del aprender haciendo. *EGA Revista de Expresión Gráfica Arquitectónica, 26*(42), 192. https://doi.org/10.4995/ega.2021.14717

Simonsen, J., & Robertson, T. (2012). *Routledge international handbook of participatory design*. Routledge.

Stebbins, R. A. (2006). Serious leisure. In C. Rojek, S. M. Shaw, & A. J. Veal (Eds.), *A handbook of leisure studies* (pp. 448–456). Palgrave Macmillan. https://doi.org/10.1057/9780230625181_27

Thornicroft, G., Mehta, N., Clement, S., Evans-Lacko, S., Doherty, M., Rose, D., et al. (2016). Evidence for effective interventions to reduce mental-health-related stigma and discrimination. *The Lancet, 387*(10023), 1123–1132.

White, M., & Epston, D. (1990). *Narrative means to therapeutic ends* (1st ed.). Norton & Company.

Advancing Inclusive Education and Enhanced Pedagogy Through Accessible Escape Rooms

Ioannis Poultourtzidis, Panagiotis Antoniou, Maria Nikolaidou, Eleni Dafli, Nikolaos Kyriakidis, Savvas Anastasiades, and Panagiotis Bamidis

Abstract

There has been increased emphasis on using immersive and interactive learning experiences to improve educational outcomes recently. Drawing from three completed research projects in medical education and digital accessibility, our goal is to create a blueprint for accessible educational escape rooms that address the unique needs of learners. We aim to design engaging and inclusive experiences that promote participation for all students by integrating insights and methods. This study builds on three successfully completed projects: developing medical education escape rooms, digitizing them for remote learning and incorporating digital accessibility tools. By utilizing expertise from education, technology and accessibility, we explore participatory design approaches and involving individuals with disabilities in the creation process. Quantitative and qualitative methods are proposed for assessing the effectiveness and accessibility of the escape rooms. Combining findings from past projects reveals that immersive educational experiences like escape rooms promote active learning and knowledge retention. Transforming physical escape rooms into digital formats supports remote education, especially in times of crisis, while evaluating accessibility tools highlights the importance of creating inclusive learning environments. Building on such insights the focus is on developing accessible escape rooms that cater to the needs of diverse learners, particularly those with disabilities. This research promotes accessible education through innovative and inclusive learning experiences, such as accessible educational escape rooms that offer a valuable approach to enhancing engagement and learning. The perspectives of people with disabilities should be considered from the design phase to ensure that the resources are more effective and user-friendly. Future research will focus on improving these escape rooms and evaluate their impact on learning outcomes and accessibility.

I. Poultourtzidis (✉) · P. Antoniou · M. Nikolaidou · E. Dafli · N. Kyriakidis · S. Anastasiades · P. Bamidis
Aristotle University of Thessaloniki, Thessaloniki, Greece
e-mail: giannis.poultourtzidis@gmail.com; lncs@springer.com

P. Antoniou
e-mail: pantonio@otenet.gr

M. Nikolaidou
e-mail: mnikola@auth.gr

E. Dafli
e-mail: elendaf@auth.gr

S. Anastasiades
e-mail: sanastas@auth.gr

P. Bamidis
e-mail: bamidis@auth.gr

S. Camplone and C. Porfirione (eds.), *Design for All and the Future of the Planet*, Design For Inclusion 6, https://doi.org/10.1007/978-3-032-21769-1_9

Keywords

Educational escape rooms · Inclusiveness · Accessibility · Active learning · Participatory design

1 Introduction

In recent times, education has shifted towards creating more engaging and hands-on learning experiences, such as escape rooms. Game-based methods are used in healthcare, through which immersive, real-life scenarios can be safely encountered. These reflect challenges that healthcare professionals face to help boost engagement, foster collaboration, and encourage critical thinking, all within a supportive, student-centered environment (Antoniou et al., 2024). Game-based learning has gained traction for its ability to enhance engagement, encourage collaboration and academic performance (Sidekerskienė & Damaševičius, 2023). Escape rooms provide unique opportunities to simulate healthcare protocols, fostering knowledge retention and skill development. Guckian found that medical students participating in escape rooms showed improved practical skills, including communication, task delegation, and leadership (Guckian et al., 2020), while others claim that escape rooms promote knowledge acquisition (Cotner et al., 2018), provide immersive activities where students demonstrate their capabilities (Clarke et al., 2017) and foster engagement (Hellín et al., 2023). These are increasingly acknowledged by educators as tools to help them achieve their goals (Veldkamp et al., 2020). The collaborative nature of escape rooms highlights the value of teamwork and dialogue, essential components in medical practice (Kinio et al., 2019). Moreover, immersive activities that simulate disabilities can cultivate empathy, preparing students to provide inclusive and compassionate care in future practice (Kletenik & Adler, 2022), while triggering active participation. This also caters for millennials' preference of activities that foster connecting to each other and encourage active participation (Jambhekar et al., 2020).

Despite the growing popularity of escape rooms, their accessibility for learners with disabilities remains underexplored. Addressing this gap is crucial, as inclusive medical education equips future healthcare professionals with the cultural competency needed to serve diverse populations and improve patient care and health outcomes (LeBlanc et al., 2020; Verbree et al., 2023). Exposure to various perspectives during training fosters empathy and cultural sensitivity, vital for building trust and reducing health disparities (George & Sidhu, 2023; McDade & Mason, 2021). By incorporating accessibility features, escape rooms can ensure equitable participation, supporting both the educational goals of the curriculum and the diverse needs of learners (Verbree et al., 2023; Dutta, 2021). Supporting diversity in residency programs, aligning providers with patients' demographics through co-created curricula and creating inclusive medical training environments assist in addressing health disparities and enhance community health (Dutta, 2021; Crites et al., 2022; Razack & Philibert, 2019). Also, it would be beneficial to tackle challenges such as outdated admission standards for underrepresented populations and create more inclusive settings through targeted recruitment efforts and inclusive strategies (LeBlanc et al., 2020; Crites et al., 2022; DeLisa & Lindenthal, 2012).

This study synthesizes insights from three successfully completed projects, namely, Escape4Health,[1] CLUEDUP,[2] and LIVE IT[3] – to create, utilizing participatory design, accessible, inclusive educational escape rooms, which according to AMEE Guide No. 168 are designed around live-action, team-based scenarios, incorporating puzzle-solving, objective-driven themes, and time-sensitive challenges. Inclusive is an intervention that can be completed without external assistance (Dudley et al., 2023). By leveraging participatory design methods that

[1] https://escape4health.eu/.

[2] http://cluedup.eu/.

[3] https://liveit-project.eu/.

involve individuals with disabilities, these escape rooms aim to offer inclusive, immersive learning experiences tailored to diverse physical, sensory, and cognitive requirements. Such initiatives seek to break down barriers in education, providing equitable opportunities for all learners to offer dynamic simulations for students to apply knowledge and find solutions (Kim et al., 2024), enhance information retention (Antoniou et al., 2024; Papamalis et al., 2022) and promote excitement and effectiveness of education experience (Nadeem et al., 2023; Ninaus et al., 2019).

2 Methods

This study synthesizes methodologies from CLUEDUP, Escape4Health, and LIVE IT to develop accessible educational escape rooms through mixed-method approaches. CLUEDUP combined game-based learning with methods like Problem-Based Learning (PBL) and Team-Based Learning (TBL) in escape room scenarios, fostering collaboration and decision-making. Stakeholder feedback gathered via interviews, surveys, and focus groups informed its iterative design. Dissemination included workshops, MOOCs, and shared insights (Papamalis et al., 2022). Escape4Health engaged stakeholders to design medical escape rooms and produced a reference guide for escape room designs for Medical Education Practioners (Fides-Valero et al., 2024). Pilot testing and feedback from post-graduate students and educators refined these designs (Bray et al., 2023). LIVE IT focused on digital accessibility for individuals with cognitive disabilities using co-design labs, interviews, and hackathons to test tools and develop an Open Toolkit (Poultourtzidis et al., 2022).

Regarding data collection, CLUEDUP assessed medical students' skills and knowledge through pre/post-training surveys, observational notes, and readiness tests. Escape4Health conducted discovery sessions and pilot testing with surveys and focus groups involving healthcare professionals and educators. LIVE IT used co-design labs, interviews, and hackathons to test digital tools, gathering data through surveys and observational analysis (Poultourtzidis et al., 2022). CLUEDUP applied statistical tests and thematic analysis to evaluate escape pedagogies' impact on learning to conduct data analysis. Escape4Health analyzed surveys and focus group data for feasibility and effectiveness of actual escape rooms in medical education. LIVE IT categorized user feedback on tools for usability, layout, and functionality (Poultourtzidis et al., 2022). CLUEDUP involved 90 medical students from AUTH, SGUL, and MU (Antoniou et al., 2024). Escape4Health engaged 275 students, 175 educators, and healthcare professionals. LIVE IT included over 100 individuals with cognitive disabilities and 35 experts contributing to the Open Toolkit's co-creation (Poultourtzidis et al., 2022). The abovementioned methods are summarized and presented to Table 1.

These projects produced a wealth of documentation both in the form of primary data (questionnaires, interviews, etc.) and in the form of aggregated data (deliverables, recommendations, academic papers). Our effort for compiling and combining these results follows an approach of theme distillation on document analysis evidence. Document analysis is a well-established qualitative research methodology, pertaining to the review and information extraction of existing documents and overall written evidence base (Bowen, 2009). Specifically, we reviewed the deliverables and publications based on the three projects and we distilled their final outcomes on three axes: Escape room design principles; Pedagogical considerations for immersive escape-like activities; Digital inclusivity considerations. From this starting point we conducted a thematic synthesis putting together initial guidelines for designing educational inclusive escape activities for medical curricula (Bowen, 2009).

Table 1 Summary of methodologies used in the three projects

Project	Methodologies	Data collection	Data analysis	Participants
CLUEDUP	Stakeholder feedback (interviews, surveys, focus groups), iterative design, MOOC, workshops	Pre/post-training surveys, observational notes, individual and team readiness tests	Statistical tests, thematic analysis to evaluate impact	90 medical students (from AUTH, SGUL, MU)
Escape4 Health	Stakeholder engagement, pilot testing, reference guide development for medical education practitioners	Discovery sessions, pilot testing with surveys, focus groups with postgraduate students and educators	Surveys and focus group analysis for feasibility and effectiveness	275 students, 175 educators, healthcare professionals
LIVE IT	Co-design labs, hackathons, interviews, digital accessibility testing	Surveys, observational analysis from pilot tests	Categorization of user feedback on usability, layout, and functionality	100+ individuals with cognitive disabilities, 35 experts contributing to the open toolkit

3 Results

What follows is the distillation of specific recommendations from each project's document base.

Escape4health provided a set of recommendations to create an escape room successfully as presented below:

E 1. Set up a design team with clearly defined roles–2 to 4 people that cover a minimum set of roles (medical, educational, gaming, participant)

E 2. Identify the audience–choose time limitations, participants' number and language style depending on age, profession and work schedule

E 3. Identify learning goals–align the activity with the healthcare curricula it supports. Learning goals can be sorted to (i) cognitive/social skills, (ii) medical knowledge and (iii) technical skills

E 4. Identify existing constraints:
- the nature of the escape rooms, such as physical of digital
- the duration of the activity: activity scope presentation a minimum of 10 min, briefing a maximum of 10 min, activity between 30 and 60 min, debriefing 30 to 45 min, assessment interviews (if any) a maximum of 30 min and assessment questionnaires (if any) a maximum of 10 min
- participants' number: it is suggested to keep the total number between 4 (to have team dynamics and variation of skills, personalities and experiences) and 8 (for all to have a chance to contribute and to maintain activity control)

E 5. Establish a setup with:
- intriguing theme definition and integration of technical content focused on medicine to trigger engagement
- having an inspired narrative to enhance experience and cohesion
- using a mood board to frame how the activity should feel
- setting the layout to set up the elements and components, depending on whether it is a permanent setting, a portable one, if it has multiple stages for parallel progress, related difficulties etc.

E 6. Design the puzzles: start by reviewing existing resources, then choose and mix considering that puzzles should be challenging but fair and balanced, provide autonomy, control and feedback, connect with the general content, promote interaction and collaboration, have elements of narrative and storytelling to trigger emotions and improve usability and accessibility. An important aspect is also to include scenario lifesavers who can correct the course of action if participants are blocked (by adapting or improvising, giving clues and hints or suggesting restarts).

E 7. Validate and iterate based on the prototype and tests, adjustments, modifications and iterations. This requires a test plan, a test group, feedback and note-keeping.

E 8. Run the activity by starting with deploying the escape room, preparing the team and using an operator manual. Then, it is incorporated into the course after lecture and briefing, and before the debriefing. When running it, it is suggested to oversee progress, detect blocks and provide hints, monitor safety, ensure the conclusion of the activity and reset the room.

CLUEDUP emphasized pedagogical considerations. Specifically, for medical, escape-oriented educational activities, recommendations include:

C 1. Identify learning objectives—concepts such as medical emergency response or teaching critical thinking and decision-making

C 2. Design the scenario in alignment with the learning objectives–with the scenarios being realistic and challenging

C 3. Diverse integration into curriculum–both standalone activities and attachments to existing modules should be integrated, to maintain relevance

C 4. Implement and constantly evaluate—to enable feedback collection and proceed with the next step,

C 5. Constant iteration for continued improvements—to refine and improve effectiveness

C 6. Embed learner-centric active pedagogical approaches—foster active learning, critical thinking and collaboration

C 7. Plan and allocate available resources – to support successful curricular implementation (time, technology, funding etc.).

LIVE IT recommendations for inclusivity include the following guidelines:

Design and Usability

L 1. Use simple, clear layouts with plain language, minimal text, and spaces between elements. Incorporate visual or audio aids to support understanding.

L 2. Ensure ease of navigation by guiding users on required actions and next steps.

L 3. Design tools that are free, ad-free, and accessible to all users, including those unable to pay for updates.

L 4. Develop tools that are compatible across browsers and operating systems, and ensure full accessibility from installation to optimization and use.

L 5. Eliminate errors and bugs to prevent confusion and frustration for PwCD.

Assistive Features

L 6. Integrate cross-platform buttons for text-to-speech, dictation, and live translation.

L 7. Use adjustable reader speeds, humanlike voices for text-to-speech, and features that encourage users to read along.

L 8. Include tools to summarize dense content, read lips for conversations, and blur backgrounds to improve focus on text.

L 9. Provide tools to customize profiles based on specific challenges or diagnoses, tailoring digital content and interfaces accordingly.

L 10. Create tools that remove, blur, or hide distracting media like ads, pictures, videos, or animations.

Content Accessibility

L 11. Facilitate understanding of complex content such as numerical data, dates, tables, and calculations through dedicated tools.
L 12. Ensure PDF accessibility by adhering to compliance standards and protocols.
L 13. Use audio/visual instructions alongside text-based content to enhance comprehension.

Empowerment and Independence

L 14. Provide training, tutorials, and resources to help users acquire digital skills and empower them to interact independently with tools.
L 15. Foster user's self-esteem, independence, and privacy by minimizing guidance by external facilitators (caregivers, family members, etc.)

Specialized Virtual Assistance

L 16. Develop virtual assistants that can adapt to various needs, such as reminders for memory challenges or reading content aloud with a single button press.
L 17. Integrate accessibility tools with visual and layout adjustments to create a cohesive user experience.

Inclusivity and Multilingual Support

L 18. Design tools for multilingual users and consider those who do not speak English, using plain language universally.
L 19. Address the needs of users with severe disabilities with robust, targeted solutions.

Community and Caregiver Support

L 20. Focus on digital skills acquisition for caregivers who act as mediators, providing training for both caregivers and end-users.
L 21. Ensure tools encourage self-reliance while supporting caregivers with the necessary know-how for effective mediation.

Based on these considerations the following initial design instructions have been derived by thematically linking relevant recommendations from all projects:

CEL 1. *An escape experience should holistically and inclusively train according to the defined learning objectives.* It is ideal for an escape experience to incorporate the required user training, including accessibility features, in the context of the escape experience, instead of a detached tutorial away from the experience. (E1, C1, L14)
CEL 2. *An educational escape experience should be adaptable according to target group.* An inclusive escape experience should be designed according to target group limitations and capacities. While a bespoke experience for each target group is unsustainable, a suite of embedded accessibility tools in each relevant puzzle can facilitate multifaceted inclusivity. (cf. E2, L6-L10)
CEL 3. *People with disabilities, or from vulnerable groups, should be in the design team of an escape activity.* Involving such users either in the preliminary design or in the feedback process builds experiences that instill confidence and

self-reliance in vulnerable learners that will encounter the escape learning experience (cf. E1, L15, L21)

CEL 4. *Inclusivity features should be the target of bespoke evaluation and multiple iterations for optimization.* The experience and the design of medical education escape rooms should be created from the ground up to be inclusivity aware, while verification should inform improvements until the experience is optimized (cf. E7, C5, C6, L1-L21)

CEL 5. *Caregivers should be actively implicated in the escape experience to maintain the authenticity and impact of the experience.* For target groups that require caregiver support, the escape experience should include roles and responsibilities for them to organically include them in the experience and maintain authenticity. (cf. L15, L20, L21, E6, C2)

CEL 6. *Digital enablers, virtual assistants and tools should be narratively included in the experience instead of detracting from it.* (cf. E6, L6-L10, L16-L19)

CEL 7. *Narrative elements should engage emotionally the users, thus increasing their engagement, with care to established inclusivity principles.* No educational escape experience should emotionally engage the users negatively, by attacking vulnerable groups, reducing their self-reliance, or putting them in positions of reduced agency. If such experiences are necessary for educational and narrative reasons (e.g. teaching empathy), then discussion and informed consent should be acquired by all participants (cf. L15, L21, E6, C6).

4 Discussion

4.1 Inclusive Learning Environments

Recent studies emphasize the importance of incorporating accessibility features to prevent exclusion. Key recommendations include simplifications of use and control of elements, patterns, icons or text, use of different formats to share information, use of color alternatives, ensuring user interface elements are user-friendly, accommodating hearing impairments with subtitles, symbols, captions, transcriptions or written instructions and enhancing visual accessibility with tactile components like Braille. The goal of this effort is to support the integration of such modalities and features into educational escape rooms. Additionally, it is crucial to promote awareness of these accessibility features so that users can fully utilize them, creators to align with them and policymakers to foster them.

Various strategies can be implemented to address the cognitive load that may hinder learners from achieving the desired learning outcomes. Initially, studies indicate that ensuring that the narrative aligns with instructional objectives keeps the content both relevant and engaging (Adams et al., 2012). Merging entertainment with educational content in a seamless manner can also improve the learning experience (Habgood & Ainsworth, 2011). Furthermore, adjusting the sequence of story events and incorporating unexpected occurrences can stimulate cognitive processes that aid in learning (Hoeken & Van Vliet, 2000).

Individuals in escape rooms are reported to be motivated and competitive while they connect the experience and its learnings to their future professional activities. They analyzed knowledge under pressure, communicated effectively, and demonstrated the dexterity and motor skills required for physical puzzles. Despite the effectiveness of escape rooms there is no one-size-fits-all approach; different learning objectives and student profiles require tailored goals. Innovative pedagogical activities like medical escape rooms should be adapted to meet these needs (Sidekerskienė & Damaševičius, 2023; Kletenik & Adler, 2022; Jambhekar et al., 2020; Dudley et al., 2023; Grammenos et al., 2009; Lage et al., 2000).

4.2 Design Process

There are claims that digital designers do not value accessibility and their outputs often push individuals with disabilities to constantly request assistance and rely on their peers or caregivers (Poultourtzidis et al., 2022). The AMEE Guide No. 168 emphasizes the importance of a design team incorporating medical, educational, gaming, and participant perspectives to create accessible and user-friendly escape rooms (Fides-Valero et al., 2024). Engaging end-users during the design phase is essential to address participants' feedback and ensure inclusivity. A participatory design approach involving individuals with disabilities enhances accessibility and efficacy and results in escape rooms that are more engaging for everyone (Coronado et al., 2023; Menzies, 2019).

4.3 Limitations

Despite positive results, escape rooms alone cannot fully achieve educational objectives. They should be paired with targeted instructional sessions, especially in medical training, to align with current pedagogical strategies (Fides-Valero et al., 2024). Additionally, obstacles might emerge in following the crucial experiential rules of escape room situations while simultaneously trying to enhance accessibility for participants (Mateen et al., 2023). Individuals with disabilities have not yet been actively involved in the escape rooms of the two projects, limiting the broader applicability of the findings. Future studies should prioritize designing customized and accessible medical education escape rooms that involve a diverse range of participants to help further develop these initiatives.

5 Conclusion

5.1 Summary

This study highlights the importance of accessibility and demonstrates how educational escape rooms promote active learning and knowledge retention. Furthermore, it has produced an initial set of specific inclusivity/accessibility recommendations for the design and implementation of medical escape-based educational experiences. Key in the implementation of these guidelines is the participatory design methodology. It ensures that tools and materials are designed from the ground up as user-friendly and accessible as possible since it incorporates effectively all stakeholder input. In that context, participatory educational design becomes key to enhancing equity, motivation, and student engagement through immersive learning experiences.

5.2 Future Research

Future studies should focus on widening the evidence base for accessibility design of educational escape rooms. The first step is the actual implementation and piloting of an inclusive design process and its resulting escape-based medical education activity. Furthermore, exploring how inclusivity features in escape rooms impact learning outcomes and engagement over time is crucial to better understand and improve their educational value. Future research can contribute to the development of innovative and widely accessible teaching techniques that are beneficial to all students by expanding current efforts.

References

Adams, D. M., Mayer, R. E., MacNamara, A., Koenig, A., & Wainess, R. (2012). Narrative games for learning: Testing the discovery and narrative hypotheses. *Journal of Educational Psychology, 104*(1), 235–249. https://doi.org/10.1037/a0025595

Antoniou, P. E., Papamalis, F., Dafli, E., Poultourtzidis, I., Schwarz, D., Woodham, L., Dimitriadis, S., Tagaras, K., Kyriakidis, N., David, P., Nikolaidou, M., Skřišovská, T., Poulton, T., & Bamidis, P. D. (2024). Medical education escape room aligned with flipped classroom and powered by Mobile augmented reality. *Electronics, 13*(12), 2367. https://doi.org/10.3390/electronics13122367

Bowen, G. A. (2009). Document analysis as a qualitative research method. *Qualitative Research Journal, 9*(2), 27–40. https://doi.org/10.3316/QRJ0902027

Bray, L., Antoniou, P., Nikolaidou, M., Fides-Valero, Á., Roberts, T., Ahonen, O., Häyrinen, S., Campos-Díaz, E., & Bamidis, P. D. (2023). Evaluation of the effectiveness of educational escape rooms within health professions education: A systematic review protocol. *International Journal of Educational Research, 117*, 102123. https://doi.org/10.1016/j.ijer.2022.102123

Clarke, S. J., Peel, D. J., Arnab, S., Morini, L., Keegan, H., & Wood, O. (2017). Escap ED: A framework for creating educational escape rooms and interactive games to for higher/further education. *International Journal of Serious Games, 4*(3). https://doi.org/10.17083/ijsg.v4i3.180

Coronado, A., Carvalho, S., & Berretta, L. (2023). Game accessibility: Adaptation of a digital escape room game to improve spatial cognitive skills on blind people. *Symposium on Virtual and Augmented Reality*, 174–182. https://doi.org/10.1145/3625008.3625039

Cotner, S., Smith, K. M., Simpson, L., Burgess, D. S., & Cain, J. (2018). Incorporating an "escape room" game Design in Infectious Diseases Instruction. Open forum. *Infectious Diseases, 5*(suppl_1), S401–S401. https://doi.org/10.1093/ofid/ofy210.1144

Crites, K., Johnson, J., Scott, N., Shanks, A., Crites, K., Johnson, J. D., Scott, N., & Shanks, A. (2022). Increasing diversity in residency training programs. *Cureus, 14*(6). https://doi.org/10.7759/cureus.25962

DeLisa, J. A., & Lindenthal, J. J. (2012). Commentary: Reflections on diversity and inclusion in medical education. *Academic Medicine, 87*(11), 1461. https://doi.org/10.1097/ACM.0b013e31826b048c

Dudley, J., Yin, L., Garaj, V., & Kristensson, P. O. (2023). Inclusive immersion: A review of efforts to improve accessibility in virtual reality, augmented reality and the metaverse. *Virtual Reality, 27*(4), 2989–3020. https://doi.org/10.1007/s10055-023-00850-8

Dutta, P. (2021). *Promoting awareness, diversity, &; inclusivity: Accessibility in medical education.* My University. https://doi.org/10.7302/1683

Fides-Valero, A., Bray, L., Dieckmann, P., Antoniou, P., Lahtinen, P., Bamidis, P., & Nikolaidou, M. (2024). Creating and running an escape room for healthcare curricula: AMEE guide no. 168. *Medical Teacher, 47*(1), 6–17. https://doi.org/10.1080/0142159X.2024.2330575

George, R. E., & Sidhu, M. S. (2023). Promoting inclusivity in health professions education. *The Clinical Teacher, 20*(6), e13606. https://doi.org/10.1111/tct.13606

Grammenos, D., Savidis, A., & Stephanidis, C. (2009). Designing universally accessible games. *Computers in Entertainment, 7*(1), 1–29. https://doi.org/10.1145/1486508.1486516

Guckian, J., Eveson, L., & May, H. (2020). The great escape? The rise of the escape room in medical education. *Future Healthcare Journal, 7*(2), 112–115. https://doi.org/10.7861/fhj.2020-0032

Habgood, M. P. J., & Ainsworth, S. E. (2011). Motivating children to learn effectively: Exploring the value of intrinsic integration in educational games. *Journal of the Learning Sciences, 20*(2), 169–206. https://doi.org/10.1080/10508406.2010.508029

Hellín, C. J., Calles-Esteban, F., Valledor, A., Gómez, J., Otón-Tortosa, S., & Tayebi, A. (2023). Enhancing student motivation and engagement through a gamified learning environment. *Sustainability, 15*(19), 19. https://doi.org/10.3390/su151914119

Hoeken, H., & Van Vliet, M. (2000). Suspense, curiosity, and surprise: How discourse structure influences the affective and cognitive processing of a story. *Poetics, 27*(4), 277–286. https://doi.org/10.1016/S0304-422X(99)00021-2

Jambhekar, K., Pahls, R. P., & Deloney, L. A. (2020). Benefits of an escape room as a novel educational activity for radiology residents. *Academic Radiology, 27*(2), 276–283. https://doi.org/10.1016/j.acra.2019.04.021

Kim, C., Zhang, N., & Bai, C. (2024). Escape rooms for education: A meta-analysis. *International Journal of Instruction, 17*(4), 219–234. https://doi.org/10.29333/iji.2024.17413a

Kinio, A. E., Dufresne, L., Brandys, T., & Jetty, P. (2019). Break out of the classroom: The use of escape rooms as an alternative teaching strategy in surgical education. *Journal of Surgical Education, 76*(1), 134–139. https://doi.org/10.1016/j.jsurg.2018.06.030

Kletenik, D., & Adler, R. F. (2022). Let's play: Increasing accessibility awareness and empathy through games. In *Proceedings of the 53rd ACM technical symposium on computer science education* (pp. 182–188). https://doi.org/10.1145/3478431.3499277

Lage, M. J., Platt, G. J., & Treglia, M. (2000). Inverting the classroom: A gateway to creating an inclusive learning environment. *The Journal of Economic Education, 31*(1), 30–43. https://doi.org/10.2307/1183338

LeBlanc, C., Sonnenberg, L. K., King, S., & Busari, J. (2020). Medical education leadership: From diversity to inclusivity. *GMS journal for Medical Education, 37*(2), Doc18. https://doi.org/10.3205/zma001311.

Mateen, S., Wiesemüller, P., & Voß-Nakkour, S. (2023, April). Design of an Accessible VR-escape room for accessibility education.
McDade, W., & Mason, B. S. (2021). Making the case for equity in graduate medical education. *Journal of the National Medical Association, 113*(1), 3–5. https://doi.org/10.1016/j.jnma.2021.01.004
Menzies, R. (2019). Unlocking accessible escape rooms. In *The 21st international ACM SIGACCESS conference on computers and accessibility* (pp. 510–512). https://doi.org/10.1145/3308561.3354602
Nadeem, M., Oroszlanyova, M., & Farag, W. (2023). Effect of digital game-based learning on student engagement and motivation. *Computers, 12*(9), 177. https://doi.org/10.3390/computers12090177
Ninaus, M., Greipl, S., Kiili, K., Lindstedt, A., Huber, S., Klein, E., Karnath, H.-O., & Moeller, K. (2019). Increased emotional engagement in game-based learning—A machine learning approach on facial emotion detection data. *Computers & Education, 142*, 103641. https://doi.org/10.1016/j.compedu.2019.103641
Papamalis, F., Antoniou, P., Poultourtzidis, I., Nikolaidou, M., Poulton, T. A., & Bamidis, P. (2022). Collaborative learning using escape designs and pedagogies. A needs analysis protocol. In *Advances in informatics, management and Technology in Healthcare* (pp. 426–429). IOS Press. https://doi.org/10.3233/SHTI220756
Poultourtzidis, I., Marina Katsouli, A., Anastasiades, S., Makroglou, S., Sidiropoulos, E., & Bamidis, P. D. (2022). Supporting digital inclusion and web accessibility for people with cognitive disabilities. In B. Séroussi, P. Weber, F. Dhombres, C. Grouin, J.-D. Liebe, S. Pelayo, A. Pinna, B. Rance, L. Sacchi, A. Ugon, A. Benis, & P. Gallos (Eds.), *Studies in health technology and informatics*. IOS Press. https://doi.org/10.3233/SHTI220543
Razack, S., & Philibert, I. (2019). Inclusion in the clinical learning environment: Building the conditions for diverse human flourishing. *Medical Teacher, 41*(4), 380–384. https://doi.org/10.1080/0142159X.2019.1566600
Sidekerskienė, T., & Damaševičius, R. (2023). Out-of-the-box learning: Digital escape rooms as a metaphor for breaking down barriers in STEM education. *Sustainability, 15*(9), 9. https://doi.org/10.3390/su15097393
Veldkamp, A., Van De Grint, L., Knippels, M.-C. P. J., & Van Joolingen, W. R. (2020). Escape education: A systematic review on escape rooms in education. *Educational Research Review, 31*, 100364. https://doi.org/10.1016/j.edurev.2020.100364
Verbree, A.-R., Isik, U., Janssen, J., & Dilaver, G. (2023). Inclusion and diversity within medical education: A focus group study of students' experiences. *BMC Medical Education, 23*(1), 61. https://doi.org/10.1186/s12909-023-04036-3

Multisensory Experience in Cultural Heritage Museums: Experimentation with More Inclusive Itineraries in Florentine Museums

Francesca Tosi, Ester Iacono, and Alessia Brischetto

Abstract

Museums, traditionally designed to preserve cultural heritage, have evolved to meet the needs of a changing society. Despite progress in removing physical barriers, many visitors still face challenges related to sensory, cognitive, and emotional needs. Current initiatives often focus on specific physical or sensory disabilities, neglecting a broader, integrated approach to inclusivity. Optimizing User Experience (UX) in museums is essential to offer engaging, personalized pathways that address diverse needs. Multisensory engagement—encompassing sight, touch, hearing, smell, and taste—is crucial for enhancing learning and interaction. Research shows that sensory involvement deepens understanding and emotional connection with art. The XALL project, conducted in three museums in Florence (Palazzo Vecchio, Museo Stefano Bardini, Museo Nazionale del Bargello), aims to make cultural heritage accessible to all, using Inclusive Design and Human-Centred Design principles. A participatory approach engaged disability associations and museum professionals to create inclusive sensory solutions. The methodology included user profiling, interaction analysis, prototype design, and usability testing. The project developed multisensory stations and customizable video guides integrating tactile descriptions, audio, and olfactory stimuli. Results showed increased engagement, improved understanding, and stronger emotional responses, particularly among visually impaired visitors. Tactile stations with 3D models and textured surfaces allowed direct interaction with artworks, enriching the experience for blind or visually impaired individuals. In conclusion, the XALL project demonstrates how multisensory approaches can transform the museum experience, making it accessible and engaging for all. The integration of advanced technologies and inclusive design principles can help create cultural spaces that cater to diverse audiences, fostering a deeper connection with cultural heritage.

Keywords

Inclusive design · Human-centred design · Design for cultural heritage · Digital design · Design for all

F. Tosi · E. Iacono (✉) · A. Brischetto
Laboratory of Ergonomics & Design (LED), Department of Architecture (DIDA), University of Florence, Florence, Italy
e-mail: ester.iacono@unifi.it

F. Tosi
e-mail: francesca.tosi@unifi.it

A. Brischetto
e-mail: alessia.brischetto@unifi.it

S. Camplone and C. Porfirione (eds.), *Design for All and the Future of the Planet*, Design For Inclusion 6, https://doi.org/10.1007/978-3-032-21769-1_10

1 Introduction

Museums, originally conceived to preserve and enhance cultural heritage, have undergone profound transformations to adapt to the needs of a constantly changing society. In recent decades, the theme of inclusion has taken centre stage, recognising access to culture as a universal right. However, despite progress in eliminating physical barriers, many visitors still face challenges related to sensory, cognitive, and emotional needs. Current initiatives often focus on specific physical or sensory disabilities, such as visual, auditory, or motor impairments, overlooking a broader and more integrated approach to inclusivity. To overcome these limitations, optimising User Experience (UX) in museums is essential. An effective museum experience must go beyond basic accessibility, offering intuitive, engaging, and personalized pathways capable of meeting the diverse needs of visitors. Multisensory engagement, understood as the integration of stimuli that engage all senses, represents a key element in fostering learning and interaction. As Falk and Dierking (1992, 2018) highlighted, sensory involvement enhances understanding and retention of exhibition content. This is influenced by situational factors, such as sociocultural, physical and visitor context, which impact how visitors engage with exhibits. When these factors are optimized, they enhance sensory experiences and facilitate deeper connections with the content.

The integration of multiple senses—sight, touch, hearing, smell, and even taste—has been recognised as an effective strategy for building meaning in the museum experience (Christidou & Pierroux, 2019; Harada et al., 2018; Levent & Pascual-Leone, 2014), moving beyond the exclusive use of sight, which has historically dominated museums (Pallasmaa, 2024; Howes, 2014). For example, tactile routes for blind and visually impaired visitors, immersive audio guides, and the use of innovative technologies are transforming the museum experience into something more inclusive and engaging. Studies such as those by Bacci and Melcher (2011) and Luo et al. (2024) emphasise that a rich sensory interaction not only improves the understanding of artworks but also stimulates personal and emotional connections with the displayed objects.

Howes (2014) emphasised the return of touch in contemporary museums, after centuries during which it had been marginalised, and reintroduced senses such as smell and taste, often ignored in traditional museum settings. In the 17th and 18th centuries, tactile interaction with objects was everyday and served scientific, aesthetic, and therapeutic purposes. However, by the nineteenth century, touch was progressively excluded, relegating visitors to passive observers. Today, programs like the British Museum's "hands-on desks" demonstrate that multisensory engagement can enrich understanding and create a deeper emotional connection with cultural objects.

A key role is played by digital technologies, such as haptic interfaces and multimedia installations, which facilitate sensory exploration. Howes stresses that these innovations transform museums into participatory spaces where visitors can actively engage their senses and experience personalized interactions. Technologies such as Augmented Reality (AR), Virtual Reality (VR), and Extended Reality (XR) further amplify this transformation, offering powerful tools to make the museum experience dynamic and accessible to all. Recent research (Waern & Løvlie, 2022; Vaz et al., 2020) highlights how interactive and immersive technologies can enhance engagement and understanding, enabling more direct and participatory interaction for all visitors.

Applying UX principles in museums also aims to optimize the interactions between visitors and cultural spaces, with the goal of creating meaningful, intuitive, and memorable experiences. According to Norman (2004), good UX is based on understanding users' expectations and meeting not only their practical needs but also their emotional ones. This principle has been adapted to the museum context to design visitor pathways that are not only accessible but also engaging and easy to navigate. Research by Pekarik et al. (1999) identifies four key dimensions of the museum experience: cognitive, emotional, social, and personal. Each dimension contributes to a fuller experience, ensuring that the visit is not merely a passage between objects and explanations, but an opportunity for reflection, interaction, and emotional engagement. The integration of advanced technologies and multisensory approaches allows museums to respond to these dimensions, offering meaningful and engaging experiences for every visitor.

The *XALL* project centres on three prominent museums in Florence: *Palazzo Vecchio*, *Museo Stefano Bardini*, and *Museo Nazionale del Bargello*. By applying **Inclusive Design** (Dong et al., 2012) and **Human-Centred Design** (ISO 9241-210: 2019), the project strives to ensure that cultural heritage is accessible to all visitors, regardless of their abilities. Through participatory design, the project actively involved several disability associations—such as **ENS (Italian Association of the Deaf)**, **UICI (Italian Union for the Blind and Visually Impaired)**, and **HABILIA Onlus (Paratetraplegic and Motor Disabled Association)**—as well as museum professionals. This approach ensured that the design solutions addressed the specific needs and expectations of both museum visitors and the professionals who manage these institutions. The project aims to shift this paradigm by offering inclusive museum experiences that encompass a broad range of sensory needs, focusing on multisensory and multimodal approaches.

2 Methodology

The methodology for this project was based on a robust framework rooted in the principles of Inclusive Design and Human-Centred Design. The process was structured into distinct operational stages, each utilizing participatory and evaluative research methods to ensure that the solutions developed addressed the diverse needs of the target audience. Both qualitative and quantitative research were employed to provide comprehensive insights into how users experienced and interacted with multisensory museum environments.

2.1 Defining User Profiles

The first stage focused on identifying and understanding the diverse needs of museum's visitors, including individuals with visual, auditory, and motor disabilities, as well as the elderly and people from various cultural backgrounds. This phase was critical in ensuring that the design process was informed by the real-world needs of the target population (Dong et al., 2012). Interviews, focus groups, and discussions with representatives from disability organizations (ENS, UICI, HABILIA) facilitated the creation of detailed user profiles that guided subsequent design phases. These profiles addressed sensory limitations and considered aspects such as technological proficiency and personal preferences, following human-centred design guidelines (ISO 9241-210: 2019).

2.2 User/Museum Interaction Analysis

The second phase involved analysing existing museum interaction models, with particular attention to how users with sensory impairments engaged with exhibits. Observational studies and literature reviews on visitor behaviour in museum spaces (Brischetto et al., 2023) were conducted. This analysis aimed to understand how individuals perceived cultural heritage, especially how multisensory engagement influenced learning and emotional connections with exhibits (Falk & Dierking, 1992). Insights from this phase informed the development of sensory interventions that enhanced interactions between users and museum environments.

2.3 Design Development and Prototyping

Based on the insights from user profiles and interaction analysis, the third phase focused on creating design solutions and prototypes for multisensory interventions. These included tactile reading stations, interactive touch-screens, and audiovisual guides integrating verbal descriptions, sound effects, and tactile feedback. An iterative prototyping approach (Brown, 2009) was adopted, wherein prototypes were tested, modified, and improved through feedback from users and museum professionals. This phase heavily incorporated co-design principles, actively involving users in the development process to ensure that final solutions met their needs and expectations (Muller & Druin, 2012).

2.4 Evaluation and Testing

The fourth phase evaluated the developed interventions using usability testing and user satisfaction surveys. Evaluation methods included:

- **Usability Testing**: Observing user interactions with multisensory installations and recording metrics such as task completion time, errors, and ease of navigation. This included evaluating the effectiveness of customizable video guides in facilitating navigation through the museum spaces and its alignment with the designed visitor journey. The study identified interaction barriers and assessed whether the interventions met user needs (Nielsen, 1994).
- **User Satisfaction Surveys**: Designed to capture emotional reactions, cognitive responses, and overall satisfaction through Likert-scale items, open-ended questions, and visual analog scales (Dillman et al., 2014).
- **Semi-Structured Interviews**: Conducted with select users to explore experiences and gather detailed feedback on the installations. These interviews provided deeper insights into user perceptions and improvement suggestions (Kvale, 1996).

2.5 Data Analysis and Interpretation

The final stage analysed data from usability testing, surveys, and interviews using qualitative and quantitative methods. Qualitative data was thematically analysed to identify patterns and insights, while quantitative data was statistically evaluated to measure intervention effectiveness. This mixed-methods approach offered a comprehensive understanding of how multisensory design impacted the museum experience, particularly in terms of accessibility and engagement.

3 Results

Through a comprehensive literature review and research into the state of the art in inclusive museum design, as well as an analysis of user needs, the *XALL* project developed a set of guidelines for inclusive design in museums. These guidelines formed the foundation for the design of the following interventions:

1. **Six multisensory stations and three tactile tablets** were created in collaboration with three museums in Florence: *Palazzo Vecchio*, *Museo Stefano Bardini*, and *Museo Nazionale del Bargello*. These stations offered tactile reading of works of art, verbal descriptions, and olfactory stimuli. By engaging multiple senses, these installations aimed to make the museum experience accessible to visitors with a variety of sensory and physical needs, ensuring a more holistic approach to cultural engagement (Fig. 1).
2. **Customizable video guides** were developed through a mobile application, allowing visitors to tailor their experience to meet individual needs. These guides provided auditory and visual content that could be adjusted for various levels of engagement, ensuring that

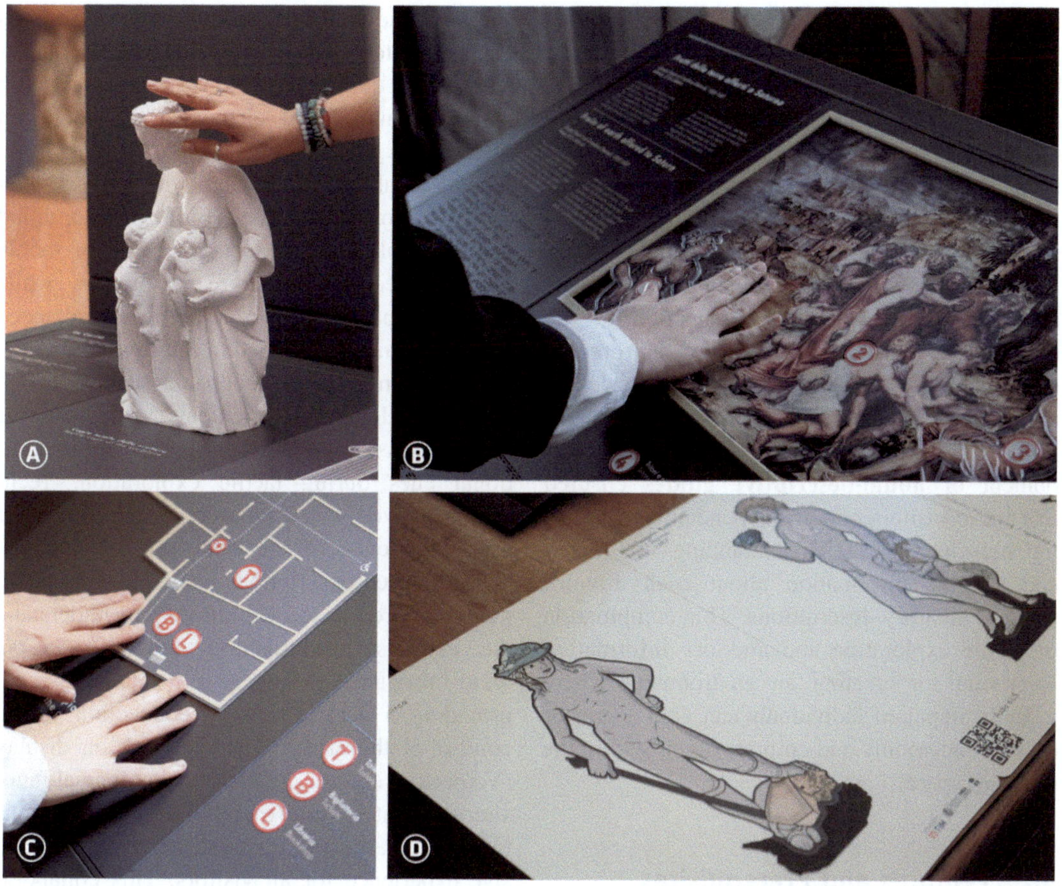

Fig. 1 Some of the multisensory stations and tactile tablets designed in collaboration with Tactile Studio. Specifically, the image illustrates some details: **a** 3D tactile reproduction of the work "Charity" inside the lapidary room of the Stefano Bardini museum; **b** 2D tactile reproduction of the work "The fruits of the earth offered to Saturn" inside the elements room of the Palazzo Vecchio Museum; **c** 2D reproduction of the tactile map to orient oneself inside the Stefano Bardini museum; **d** tactile tablets depicting "Bacchus" by Michelangelo Buonarroti and "David" by Donatello inside the Bargello National Museum

people with different sensory abilities could explore the museums in a way that suited them best.

3.1 Focus on the Tactile Dimension for Visually Impaired Visitors

One of the primaries aims of the *XALL* project was to enhance accessibility for visually impaired individuals, a group that is often underrepresented in cultural experiences. To achieve this, a specific focus was placed on the **tactile dimension** of the multisensory stations. Research indicates that the sense of touch can significantly enhance the understanding and appreciation of art for individuals with visual impairments (Cavazos Quero et al., 2021; Schiff & Foulke, 1982). By incorporating tactile elements into the exhibition design, such as 3D-printed replicas of sculptures and textured surfaces that mirror the visual characteristics of the artwork, the *XALL* project enabled visually impaired visitors to engage with the pieces in a more direct and meaningful way. Studies have shown that the tactile exploration of art can activate different cognitive processes, allowing for an increased understanding of the artwork's form, texture, and structure (Bleau et al., 2023). In the context of the *XALL* project, the tactile tablets and multisensory stations provided a rich, hands-on experience that was complementary to the verbal descriptions offered via audio guides. This integration of touch, sound, and smell allowed for a more immersive experience, fostering a deeper connection to the artwork and making the museum environment more inclusive. Furthermore, the project's design also addressed the need for **reliable navigation** for visually impaired visitors. The development of **audio-descriptive guides** ensured that these visitors could independently navigate the museum spaces, understanding the layout and specific information about each artwork through auditory descriptions. This combination of tactile exploration and auditory information is crucial for creating an environment where visually impaired individuals can gain a greater sense of autonomy and empowerment within the museum space.

3.2 Features and Prerequisites of Multi-sensory Stations

In phase 2 of the *XALL* research project, we focused on identifying the most effective methods for presenting artworks in ways that are both accessible and engaging for individuals with diverse sensory abilities. This phase involved analysing the optimal techniques for graphic representation, the materials and technologies used, and the construction methods that would best support the tactile, visual, and sensory dimensions of the museum experience. The project focused on designing multi-sensory stations that address different senses, particularly touch, while incorporating visual and auditory elements to create an inclusive visitor experience. Through collaboration with disability associations and museum staff, the team developed concepts based on the specific needs of individual museum routes and stages of the visitor journey (Tosi et al., 2024). A key aspect of the stations was the integration of tactile and visual elements. The inclusion of 2D and 3D models allowed visitors with sensory disabilities to engage with artworks and artifacts more effectively. For example, blind visitors could explore tactile models or tactile paths to feel the structure of objects, while those with partial sight benefited from enhanced graphic elements designed to improve contextual understanding. Haptic feedback, such as raised patterns and vibrating elements, added another layer of sensory engagement by providing additional cues during tactile exploration. Technology played a crucial role in enhancing the user experience of the multi-sensory stations. Voice-assisted descriptions and adjustable audio guides offered an accessible way to understand the exhibits. Additionally, olfactory sensors were introduced to engage the sense of smell, providing a deeper sensory connection to the context of the artifacts and artwork on display (Verbeek et al., 2022). These technological innovations went beyond traditional visual methods to ensure a more comprehensive and inclusive experience for all visitors. This collaborative and multi-faceted approach highlights the potential of multi-sensory design in making museums more accessible to diverse audiences, aligning with best practices in inclusive design.

3.2.1 The Three "Senses" of Tactile Perception

The effectiveness of tactile stations hinges on their ability to engage the **three essential "senses" of touch**, each of which plays a crucial role in how visitors interact with the objects and exhibits:

1. **Line Recognition (Sensory Recognition Capabilities of the Fingertip)** Visitors with visual impairments often rely on their fingertips to trace the outlines and details of objects. The ability to **follow a line**—whether the outline of a statue, the contour of an archi-

tectural element, or the edges of artwork—is a fundamental component of tactile interaction. This aspect of tactile perception ensures that visitors can navigate the shape and structure of an artwork by simply following its tactile representation.

2. **Haptic Ability (Positioning and Movement Awareness)** The **haptic ability** refers to the visitor's capacity to perceive and recognize the position of their hand in relation to the tactile device. It involves understanding the spatial dimensions of the object and being able to move around it with confidence. The design of multi-sensory stations must, therefore, account for the need for users to comprehend the spatial relationships of various elements within the exhibit, such as depth and orientation. This type of interaction enhances the visitor's ability to understand the full form of the artwork or artifact, even in the absence of visual input.
3. **Surface Recognition (Texture Perception)** Recognizing and differentiating the **roughness of surfaces** is another key tactile ability. By feeling variations in texture, individuals can gain information about the material properties of an object, such as whether it is rough, smooth, or intricately detailed. Texture is crucial for distinguishing between different materials and forms, which is particularly relevant when interpreting sculptures, pottery, or textiles.

4 Conclusions and Discussion

The design solutions developed in the XALL "Tutta un'altra Guida" project were implemented and tested in three major museums in Florence, with significant outcomes regarding accessibility and visitor engagement. The results demonstrated that multisensory interventions markedly enhanced the museum experience, making it more inclusive for a wider range of visitors. Key findings from the evaluation include: (1) Increased Engagement. The tactile stations and multimedia guides played a crucial role in increasing visitor engagement, particularly for individuals with sensory impairments. These features allowed users to experience artworks beyond the visual domain, fostering a more immersive and memorable visit (Tosi et al., 2024; Vikmane et al., 2024); (2) Improved Learning and Retention. Visitors reported improved understanding and retention of exhibit content when engaging with multisensory features. This aligns with the existing literature that suggests multisensory experiences promote deeper learning (Falk & Dierking, 2018); (3) Positive Emotional Responses. Visitors, especially those with visual impairments, expressed heightened emotional responses to the exhibits. The tactile experiences combined with soundscapes were cited as particularly impactful, further enhancing emotional engagement with the artwork.

Future research should focus on expanding the use of **multimodal and multisensory approaches** to accessibility, particularly in the context of museums and cultural institutions. There is a need for further experimentation with new technologies and design strategies that engage multiple senses, allowing for more dynamic and inclusive experiences for people with disabilities. Additionally, ongoing user feedback and usability testing will be essential to refine these interventions and ensure that they continue to meet the needs of diverse audiences. The results of the XALL project provide valuable insights for future initiatives aimed at improving the accessibility of cultural institutions. As museums continue to innovate and expand their offerings, creating environments where all visitors—regardless of their sensory or physical abilities—can experience and appreciate cultural heritage is essential.

Acknowledgments The authors would like to thank: the TIM Foundation, which financed the research; the polytechnic of Milan, Department of Electronics, Information and Bioengineering with professor Lanzi Pierluca and dr. Paolo Boffi subsequently developed the computer system; the Association MUS.e Musei Civici Fiorentini, Florence, Italy (Palazzo Vecchio and Stefano Bardini Museum) and the Bargello National Museum, Florence, Italy; the associations: ENS (Italian Association of the Deaf), UICI (Italian Union of the Blind and Visually Impaired) and

HABILIA Onlus (Paratetraplegic and Motor Disabled Association).

Author Contributions Conceptualization, A.B. and F.T; methodology, A.B., E.I; writing-original draft preparation, E.I.; writing-review and editing, E.I., A.B. (Writing–Introduction and Methodology: E.I.; Result and Conclusion and Discussion: E.I, A.B.). Supervision, F.T. and A.B.; project responsible, F.T.; funding acquisition, F.T. All authors have read and agreed to the published version of the manuscript.

References

Bacci, F., & Melcher, D. (Eds.). (2011). *Art and the senses.* Oxford University Press.

Bleau, M., van Acker, C., Martiniello, N., Nemargut, J. P., & Ptito, M. (2023). Cognitive map formation in the blind is enhanced by three-dimensional tactile information. *Scientific Reports, 13*(1), 9736.

Brischetto, A., Iacono, E., Becchimanzi, C., & Tosi, F. (2023). Enhancing inclusive experience in museums: Results from a field study. *AHFE International*, 132–144.

Brown, T. (2009). *Change by design: How design thinking creates new alternatives for business and society.* Collins Business.

Cavazos Quero, L., Iranzo Bartolomé, J., & Cho, J. (2021). Accessible visual artworks for blind and visually impaired people: Comparing a multimodal approach with tactile graphics. *Electronics, 10*(3), 297.

Christidou, D., & Pierroux, P. (2019). Art, touch and meaning making: An analysis of multisensory interpretation in the museum. *Museum Management and Curatorship, 34*(1), 96–115.

Dillman, D. A., Smyth, J. D., & Christian, L. M. (2014). *Internet, phone, mail, and mixed-mode surveys: The tailored design method.* John Wiley & Sons.

Dong, H., Clarkson, J., Coleman, R., & Cassim, J. (2012). *Design for inclusivity: A practical guide to accessible, innovative and user-centred design.* Gower Publishing, Ltd..

Falk, J. H., & Dierking, L. D. (1992). *The museum experience.* Whalesback Books Washington.

Falk, J. H., & Dierking, L. D. (2018). *Learning from museums.* Rowman & Littlefield.

Harada, T., Hideyoshi, Y., Gressier-Soudan, E., & Jean, C. (2018). Museum experience design based on multisensory transformation approach. In *DS 92: Proceedings of the DESIGN 2018 15th international design conference* (pp. 2221–2228).

Howes, D. (2014). Introduction to sensory museology. *The Senses and Society, 9*(3), 259–267.

ISO 9241-210. (2019). *Ergonomics of human-system interaction-human-centered Design for Interactive Systems.* International Organization for Standardization.

Kvale, S. (1996). *InterViews: An introduction to qualitive research interviewing.* Sage.

Levent, N., & Pascual-Leone, A. (Eds.). (2014). *The multisensory museum: Cross-disciplinary perspectives on touch, sound, smell, memory, and space.* Rowman & Littlefield.

Luo, D., Doucé, L., & Nys, K. (2024). Multisensory museum experience: An integrative view and future research directions. *Museum Management and Curatorship*, 1–28.

Muller, M. J., & Druin, A. (2012). Participatory design: The third space in human–computer interaction. In *Human computer interaction handbook* (pp. 1125–1153). CRC Press.

Nielsen, J. (1994). *Usability engineering.* Morgan Kaufmann.

Norman, D. (2004). *Emotional design: Why we love (or hate) everyday things.* Basic books.

Pallasmaa, J. (2024). *The eyes of the skin: Architecture and the senses.* John Wiley & Sons.

Pekarik, A. J., Doering, Z. D., & Karns, D. A. (1999). Exploring satisfying experiences in museums. *Curator: The Museum Journal, 42*(2), 152–173.

Schiff, W., & Foulke, E. (Eds.). (1982). *Tactual perception: A sourcebook.* Cambridge University Press.

Tosi, F., Brischetto, A., Iacono, E., & Becchimanzi, C. (2024). XALL-Tutta un'altra guida. Design per l'inclusione: strategie e strumenti per musei inclusivi. In *Design per la Diversità. Atti della Conferenza Nazionale SID 2023* (pp. 270–280). SID Società Italiana Design.

Vaz, R., Freitas, D., & Coelho, A. (2020). Blind and visually impaired visitors' experiences in museums: Increasing accessibility through assistive technologies. *The International Journal of the Inclusive Museum, 13*(2), 57.

Verbeek, C., Leemans, I., & Fleming, B. (2022). How can scents enhance the impact of guided museum tours? Towards an impact approach for olfactory museology. *The Senses and Society, 17*(3), 315–342.

Vikmane, E., Ņikitina, M., Brutāne, L., & Cērpa, L. K. (2024). Multisensory approach to museum accessibility and experience enhancement. *Culture Crossroads, 25*, 21–32.

Waern, A., & Løvlie, A. S. (2022). *Hybrid museum experiences: Theory and design* (p. 198). Amsterdam University Press.

Active Ageing: Can Design Ensure Cultural Participation From Home?

Isabel Leggiero, Francesca Rocca, and Elena Polleri

Abstract

The rapid growth of the ageing population calls for innovative strategies to ensure well-being, particularly for housebound. This research investigates the relationship between cultural participation and psychological, physical, and social health, emphasizing its role in promoting well-being and preserving intangible cultural heritage, such as oral traditions, arts, and personal stories. Using a participatory design approach, the study involves older adults co-creating inclusive digital solutions that enable cultural engagement from home. Through methodologies like living labs, simulations, and ethnographic interviews, the research explores preferences for communication channels and digital tools, uncovering barriers and motivations for participation. This study contributes to the broader discourse on cultural welfare and digital inclusion, offering a framework for designing tools that empower users, preserve collective memory, and promote social equity. Bridging the gap between cultural participation and well-being, the research provides actionable insights for creating sustainable, human-centered solutions that enhance the quality of life for older adults and beyond.

Keywords

Participatory design · Intangible cultural heritage · Cultural participation · Digital inclusion · Ageing population

I. Leggiero (✉) · F. Rocca
Università della Campania 'L. Vanvitelli', Aversa, Italy
e-mail: isabel.leggiero@unicampania.it

F. Rocca
e-mail: francesca.rocca@unicampania.it

E. Polleri
Università di Genova, Genoa, Italy
e-mail: elena.polleri@edu.unige.it

1 Research Context and Objectives

In recent years, the link between cultural participation, psychological well-being and health has gained greater attention, becoming a prominent focus in both research and local social-health policies (Fancourt & Finn, 2019). This growing interest translates into an increase in scientific studies and experimental initiatives by local administrations and social-health structures.

Traditionally, approaches aimed at improving quality of life and services focused solely on health, neglecting the socio-relational dimension of care and limiting support to technical assistance.

The introduction of cultural participation in care and non-care environments could therefore represent an important innovation, promoting more humane practices in well-being

S. Camplone and C. Porfirione (eds.), *Design for All and the Future of the Planet*,
Design For Inclusion 6, https://doi.org/10.1007/978-3-032-21769-1_11

by addressing cognitive and emotional dimensions (Sacco, 2018).

Recent neuroscientific research has shown that cultural experiences can evoke complex and profound responses, suggesting that experiential care capable of including culture is effective in ensuring a good quality of life. Furthermore, cultural participation can play a preventive role, helping to prevent pathogenic behaviors through the promotion of positive behavioral responses (Dolcos, Moore, & Katsumi, 2018).

We are at the forefront of a renewed approach to well-being, integrating physical health care with prevention and cultural welfare, a multidisciplinary practice that studies and enhances the relationship between culture and health to promote the positive effect of culture and art on individual and collective well-being in a perspective of social equity and sustainable development (Zbranca et al., 2022; Sacco, 2023).

To realize this potential, constructive dialogue between cultural and medical-healthcare sectors is essential, addressing challenges related to communication, prejudice, and legitimacy. These two fields must work together to establish relationships of trust and mutual recognition, essential for effective collaboration.

In this regard, in recent years the European Commission has published the New Agenda for Culture, a key document that will guide European cultural policies in the near future. The underlying objective of this agenda is the interaction between culture and sectors related to well-being, encouraging the European Union to invest significant resources in scientific research, clinical trials, and projects in this area.

This favorable context presupposes that it is the right time to launch initiatives capable of suggesting new formulas to ensure social well-being (European Commission, 2009; 2019).

This research aims to ensure well-being through cultural participation at home, creating a service that can improve the quality of life on an emotional, mental, physical, social, and spiritual level. At the same time, it wants to ensure that the intangible cultural heritage, which includes living traditions such as oral expressions, performing arts, social practices, rituals, celebrations, knowledge of nature and traditional crafts, is not lost. This heritage is essential to keep cultural diversity alive in a globalized world, promoting intercultural dialogue and mutual respect.

As already reported above, several studies have shown that sharing and listening to cultural content can significantly improve health and well-being conditions. However, to date, there are few tools capable of effectively reaching a large audience, often resulting in services intended for a narrow target.

An example of cultural participation that has developed in recent years is the living library, an innovative cultural initiative that offers people the opportunity to listen to life stories told in the first person by individuals, called for the occasion "living books". In this context, books are not made of paper but are instead real people who preserve unique and significant stories to share. The concept of the living library was born in Denmark in 2000, conceived by the non-governmental organization "Stop the Violence", with the idea of promoting dialogue, reducing prejudice, and fighting discrimination through direct and personal meetings with people representing various social groups often subject to negative stereotypes. Although this tool is useful in encouraging dialogue and promoting well-being in the population, it has not yet developed an adequate system to collect these testimonies and thus be able to propose them in the future and to a large audience (Policardo & Prati, 2023).

The aim of the research is, therefore, to create on the one hand a service accessible to all, therefore capable of adapting to the needs of people regardless of age, origin or physical and social condition (Longo et al., 2017), and on the other an inclusive and accessible system capable of ensuring well-being through cultural participation directly from one's home environment (Leggiero, 2024; Wiles et al., 2012).

This study focuses mainly on the elderly as an initial target, intending to then extend the results to a wider audience. The decision to base the research on an elderly sample stems from four main motivations:

- The elderly represent a considerable part of the population (Vereinte Nationen, 2023; Istat, 2023);
- They have a lot of free time to devote to alternative activities to work;
- They have a life full of experiences to tell;
- They need well-being understood as an element capable of significantly improving the quality of their life.

The objective of the present research is to develop a technological product/service that facilitates the transmission and enjoyment of personal stories, thereby ensuring the preservation and enhancement of this intangible cultural heritage. To achieve this aim, the potential for incorporating aspects related to materiality is also evaluated to foster greater empathy in the narrative. The presentation of an object associated with memory has been shown to facilitate the development of empathy in the listener, thereby enhancing their ability to empathise with the narrator and identify with the narrative to a greater extent. The implementation of such a system has the potential to effect a paradigm shift in the manner in which individuals engage with cultural phenomena, thereby fostering well-being and social inclusivity on a substantial scale.

2 Methods Used to Undertake the Research

The research is closely linked to Participatory Design, as its methodological aspect is central to achieving the desired outcomes. The use of this approach is motivated not only by the techniques employed but also by the objectives to be attained. Participatory Design (PD) is based on six key principles (Simonsen & Robertson, 2013) that have guided the definition of the objectives and activities:

- Equalizing power relations: efforts have been made to give voice to those who might otherwise remain invisible or marginalized, such as older adults who are unable to physically reach their communities, or individuals with limited economic resources and influence in local contexts.
- Democratic practices: these have been implemented to promote equality among all stakeholders involved in the decision-making process and to act in the common interest.
- Situation-based actions: the use of simulation techniques and Living Labs aims to work directly with people in their home environments to understand actions related to the use of technology in real-world contexts, rather than through formal abstractions.
- Mutual learning: opportunities for exchange and collaborative approaches have been sought to improve the understanding among various actors involved, thus facilitating the sharing of knowledge and experiences between experts and older adults.
- Tools and techniques: techniques and activities have been designed to allow stakeholders to express their needs and visions, such as knowledge-sharing sessions, brainstorming on potential prototypes, workshops, and other active engagement methods.
- Alternative visions about technology: alternative perspectives on technology use are promoted, encouraging older adults to experiment with new forms of interaction within the home environment (Mincolelli & Marchi, 2020).

To achieve the objectives that had been set out at the outset, the work is based on the use of several complementary methodologies that address the different aspects of the research. Specifically, preliminary desk research was carried out, drawing on insights from various disciplines to adopt a holistic approach to defining well-being, in line with the *One Health concept.*[1] The topics

[1] It is a holistic methodological framework that emphasizes the global health of the planet as equally important

investigated included issues related to ageing, cultural participation, and the impact of this participation on both well-being and the local economy. Concurrently, emphasis was placed on the participatory nature of the research, with an in-depth exploration of simulation techniques and Living Lab methods applied to design processes (Følstad, 2008). The involvement of users was delineated for the various phases of the research through configuring participation (Vines, Clarke, Wright, McCarthy, & Olivier, 2013), recognised as a pivotal initial design activity. At the project's inception, a sociologist specialising in ageing was involved to determine how best to explore the relationship between technology and fragile user groups, particularly older adults. Similarly, future work will require the involvement of psychologists and gerontologists to understand how to generate well-being through participatory activities conducted at home, including in cases of reduced mobility, cognitive difficulties, and related challenges.

Questionnaire

Following the research phase, a questionnaire was administered to a large audience. This was administered to over 1000 Unige Senior students (a university course dedicated to students aged 50 and above) in order to understand the preferences and usage habits regarding technological tools within the domestic environment (Marshall et al., 2022), as well as the relational and social aspects of technology. The questionnaire was divided into three sections. Firstly, respondents were asked questions about the various technological devices they used at home, with the aim of determining which devices people were most familiar with, albeit on a statistical level. Subsequently, two scenarios were presented. The first scenario simulated the conditions imposed by the pandemic, such as the inability to leave home and meet others, to understand the current use of technology as a means of social integration within the community. The second scenario concerned the sharing of culture and personal experiences.

Simulation session

In the subsequent phase of the study, a select group of students who had previously engaged with the questionnaire component was invited to partake in a preliminary simulation designed to acclimatise them to the impending final activity. The simulation entailed the selection of a personal anecdote to be shared, encompassing memories, values, and knowledge. Participants were then prompted to choose a technological device from a pre-defined list, including smartphones, laptops, and tablets. Following this, they were required to opt for a communication channel, namely voice, video, or text, to convey their narrative in their preferred modality and with the device of their selection. The facilitation strategy employed during the simulation involved a gradual shift in interaction dynamics. Initially, participants were encouraged to share their personal stories in a self-directed manner. Subsequently, in the second half of the simulation, the facilitators initiated a dialogue by posing questions that were related to the content of the stories. Concurrently, the facilitators addressed the description of multisensory elements that could serve to enhance the narrative.

The decision to utilise a simulation was substantiated by the recognition that aspects of well-being frequently elude straightforward recognition or quantification. Consequently, the capacity to simulate actions and evaluate subjective feelings proved advantageous in assessing satisfaction levels through conclusive interviews (Vemuri et al., 2020) [Images 1 and 2].

3 Research Findings

The first results were obtained from the questionnaire. A total of 240 responses were collected from participants aged 53 to 92, 73% of whom were women, with the remainder being men. The entire sample acknowledged the potential

as the health of individual beings. It therefore seeks to address issues through interdisciplinary collaboration.

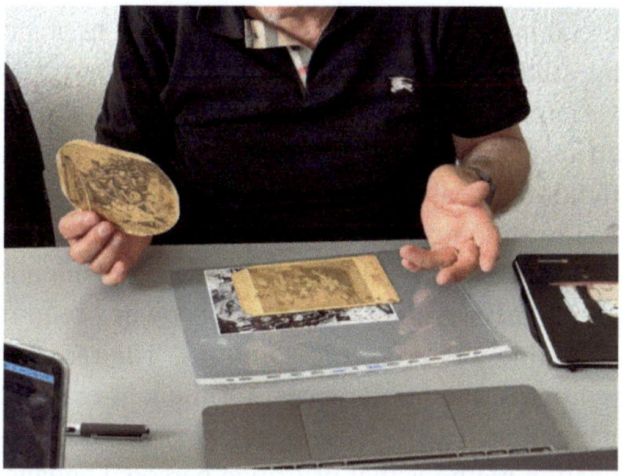

Image 1 Simulation session and interview on activity perception © the authors

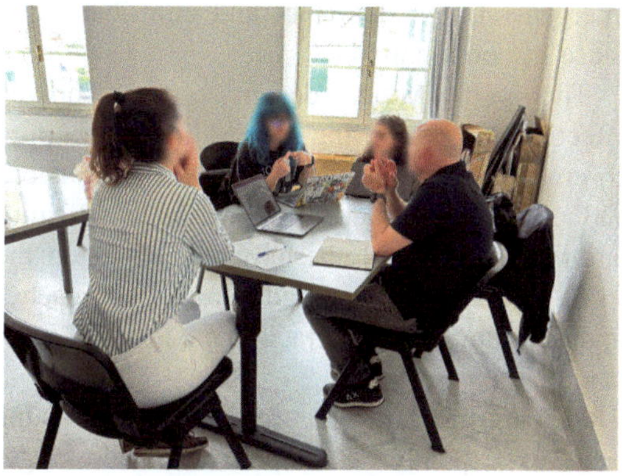

Image 2 Simulation session and interview on activity perception © the authors

of digital technology compared to analog tools, but only 31% felt they were able to fully utilize it. The most commonly used tools at home, in order of prevalence, were smartphones (99%), printers (73%), laptops (67%), scanners (55%), smart TVs (51%), tablets (40%), and desktop computers (37%).Regarding the use of technology for relational and social purposes, 43% stated they used it frequently to stay in touch with friends and family, 52% used it moderately, and 5% rarely. Participants were then presented with a scenario reminiscent of the COVID-19 quarantine period, during which leaving the home and engaging in social interactions was restricted to the essentials (Presti, 2022). In this context, 65% of respondents expressed a desire to stay connected with their community via video calls and phone calls, while 42% would also use text messages. Consequently, the most frequently used technological tools would be smartphones (94%) and laptops (49%). While 80% of respondents indicated they would have no difficulty using digital devices, 13% admitted to having little familiarity with technology, 5% expressed fear of making mistakes, and 2% found it challenging to understand how to use them.The second scenario focused on the transmission of cultural heritage. In addition to the inability to leave the home, participants were asked by the "City Museum" to share their values and knowledge with younger generations (Hovi-assad, 2016; Magliacani & Sorrentino, 2022). While 73% were willing to share their knowledge, the remaining 27% were not, citing various reasons such as: "I feel uncomfortable," "I'm shy," "I don't have skills," "I wouldn't know what to talk about," "I think I don't have interesting things to say," "I have difficulty speaking in public," or "I'm not used to talking much."

Participants who were interested in sharing expressed a desire to contribute in the following areas:

- Knowledge: professions, art, politics, architecture, music, feminist movements, childhood memories, and traditional crafts.
- Manual skills: cooking, sewing, painting, and photography.
- Values: community, lifestyles, and volunteering.
- Cultural experiences: life before modern technology, books, travel, dialects, games, memories of past generations, and traditions.

Regarding the tools used for sharing, 52% were willing to use new digital tools, 41% preferred using familiar ones, and the remaining respondents were divided between those who would not use technology at all and those who were unsure. Finally, 81% preferred speaking to real people (e.g., via calls, video calls, or online meetings), 23% favored sharing content autonomously without human mediation, and only 4% were open to interacting with interfaces simulating human interaction (e.g., chatbots, avatars, or robots). In conclusion, the motivations for sharing knowledge through technology were also investigated. The most commonly cited reasons included:

- The interest shown by listeners, opportunities to interact, receive relevant requests and feedback, and feel useful.
- Preventing the loss of testimonies from the past that could improve the future, preserving collective memory, and maintaining traditions.
- Cultural exchange and mutual enrichment, including the learning of digital tools.
- Fear of social isolation, a desire to stay connected, and a willingness to expand networks, including with people of different ages and generations.

The next activity involved seven older adults simulating the sharing of their knowledge using technology to better understand and design the final activity. The number of participants allows for only a partial analysis of the results, as it represents a limited sample.

However, regarding the facilitation of the session, participants expressed a preference for being listened to without interruptions, with questions and clarifications left until the end of the story. Furthermore, with regard to the interaction with technology, the participants expressed a preference for either speaking with a human being or recording content themselves. None of the participants indicated any interest in interacting with avatars or virtual interfaces, as these were considered to lack empathy.

4 Conclusions

This research underscores the importance of leveraging cultural participation as a tool to enhance well-being and foster inclusion for housebound individuals, particularly older adults. By exploring the intersection of design, digital inclusion, and intangible cultural heritage, the study provides valuable insights into how participatory design approaches can address the socio-relational dimensions often overlooked by traditional healthcare models.

The findings reveal that while technology holds great potential, significant barriers remain, particularly among older users who require accessible and empathic solutions. The research highlights the preferences for human-centered interactions, with participants favoring communication methods that prioritize personal engagement over virtual interfaces. Moreover, the motivation to share cultural knowledge stems from a desire to preserve collective memory, foster intergenerational dialogue, and contribute to a sense of community.

The participatory design process proved effective in empowering users, allowing them to co-create solutions that align with their needs and capabilities. Methodologies such as simulations and Living Labs have demonstrated the value of contextualized and iterative approaches in developing tools that facilitate cultural sharing from home.

Moving forward, the research emphasizes the need to expand the scope to include larger and more diverse samples, as well as to integrate multidisciplinary expertise from psychologists, gerontologists, and cultural professionals. The ultimate goal is to design scalable, inclusive, and sustainable systems that promote cultural participation, enhance well-being, and ensure the preservation of intangible cultural heritage.

By bridging the gap between technology and cultural engagement, this study offers a pathway for fostering social equity, promoting active ageing, and enriching the lives of individuals and communities alike. The results serve as a foundation for future research and innovation in designing human-centered systems for cultural and social inclusion.

Acknowledge This contribution is the result of a collaborative reflection between all the authors. The "Research context and objectives" paragraph has to be attributed to Francesca Rocca, "Methods used to undertake the research" has to be attributed to Isabel Leggiero, and "Research findings and Conclusions" to Elena Polleri.

References

Dolcos, S., Moore, M., & Katsumi, Y. (2018). Neuroscience and Well-being. In *Handbook of Well-being* (pp. 281–305). NOBA Scholar.

European Commission. (2009). Living labs for user-driven open innovation–an overview of the living labs methodology, activities and achievements.

European Commission. (2019). Quadro strategico per la politica culturale dell'UE. *Culture and Creativity*. https://culture.ec.europa.eu/it/policies/strategic-framework-for-the-eus-cultural-policy

Fancourt, D., & Finn, S. (2019). *What is the evidence on the role of the arts in improving health and Well-being?: A scoping review*. WHO Regional Office for Europe.

Følstad, A. (2008). Living labs for innovation and development of information and communication technology: A literature review. *Electronic Journal of Virtual Organization and Networks, 10*, 99–131.

Hovi-assad, P. (2016). The role of the Museum in an Ageing Society. *Museum International, 68*(3–4), 84–97. https://doi.org/10.1111/muse.12129

Istituto Nazionale di Statistica (ISTAT). (2023). Pillole Rapporto annuale 2023. https://www.istat.it/it/files/2023/07/PILLOLE-RAPANN-2023.pdf

Simonsen, J., & Robertson, T. (Eds.). (2013). *Routledge international handbook of participatory design*. Routledge. https://doi.org/10.4324/9780203108543

Leggiero, I.. (2024). Cultural participation from home: A research proposal for inclusive design. In *International conference IDEA–investigating Design in Architecture*. Genova University Press. ISBN: 978–88–3618-261-9.

Longo, A., Rabbiosi, C., & Salvadeo, P. (A c. Di). (2017). Forme dell'inclusività: Pratiche, spazi, progetti. Maggioli editore.

Magliacani, M., & Sorrentino, D. (2022). Reinterpreting museums' intended experience during the COVID-19 pandemic: Insights from Italian university museums. *Museum Management and Curatorship, 37*(4), 353–367. https://doi.org/10.1080/09647775.2021.1954984

Marshall, B. L., Dalmer, N. K., Katz, S., Loos, E., Gómez, D. L., & Peine, A. (2022). Digitization of aging-in-place: An international comparison of the value-framing of new technologies. *Societies, 12*, 35. https://doi.org/10.3390/soc12020035

Mincolelli, G., & Marchi, M. (2020). Inclusive participation design methodologies for digital cultural heritage. In G. Di Bucchianico (A c. Di), *Advances in Design for Inclusion* (Vol. 954, pp. 271–281). Springer International Publishing. https://doi.org/10.1007/978-3-030-20444-0_26

Policardo, G., & Prati, F. (2023). Don't judge a person by its cover: Human library un intervento efficace per ridurre i pregiudizi?. The inquisitive. *Mind, 23*, 1–6.

Presti, O. L. (2022). Covid-19 and the cultural life of older people. *Museum Management and Curatorship, 37*(4), 342–352.

Sacco, P. (2018). *Partecipazione culturale, benessere psicologico e salute: verso un nuovo modello di community welfare a base culturale*. Fondazione Compagnia San Paolo.

Sacco, P. (2023). *Cultura e salute: verso un modello di welfare culturale*. Università degli Studi "G. d'Annunzio" Chieti-Pescara.

Simonsen, J., & Robertson, T. (Eds.). (2012). *Routledge International Handbook of Participatory Design* (1st ed.). Routledge. https://doi.org/10.4324/9780203108543

Vemuri, S., Hynson, J., Gillam, L., & Williams, K. (2020). Simulation-based research: A scoping review. *Qualitative Health Research, 30*(14), 2351–2360. https://doi.org/10.1177/1049732320946893

Vereinte Nationen (A c. Di). (2023). Leaving no one behind in an ageing world. United Nations.

Vines, J., Clarke, R., Wright, P., McCarthy, J., & Olivier, P. (2013). Configuring participation: On how we involve people in design. In *Proceedings of the ACM conference on human factors in computing systems (CHI'13)* (pp. 429–438). Association for Computing Machinery.

Wiles, J. L., Leibing, A., Guberman, N., Reeve, J., & Allen, R. E. S. (2012). The meaning of «aging in place» to older people. *The Gerontologist, 52*(3), 357–366. https://doi.org/10.1093/geront/gnr098

Zbranca, R., Dâmaso, M., Blaga, O., Kiss, K., Dascăl, M., Yakobson, D., & Popescu, O. (2022). CultureForHealth Report. Culture's contribution to health and well-being. A report on evidence and policy recommendations for Europe.

Good Practices: Spaces and Environments

Carpe Diem Dementia Village

From Institution to a Homely Garden Village

Camilla Heier Anglero

Abstract

Carpe Diem Dementia Village in Bærum, Norway, represents a landmark shift in healthcare architecture for people living with dementia. Commissioned by Bærum Municipality and designed by Nordic Office of Architecture with Landscape architects Bjørbekk & Lindheim, the project challenges the institutional nursing-home model by creating a homely, community-integrated village where residents can live safely, actively and with dignity. The 18,000 m^2 facility comprises 136 communal housing units, 22 high-care dementia units and shared public functions, including a bistro, urban square, gardens and pathways that connect the village to its surroundings. Rather than designing primarily around limitations, the project translates service-design insight, workshops and user surveys into architecture that supports individual capability, orientation and everyday choice. A small-town domestic structure, varied building heights, recognizable entrances, gardens, colours and material identities create intuitive wayfinding and a sense of belonging. Durable brick and untreated wood provide warmth, robustness and environmental quality in line with Nordic Ecolabel requirements. Completed in 2020, Carpe Diem has become a national and international reference for inclusive design, demonstrating how healthcare environments can move beyond care provision to become social, architectural and civic ecosystems for better living. It offers a transferable model for future dementia care, ageing communities and healthcare planning.

Keywords

Dementia care · Inclusive design

1 Introduction

> We all strive to live good lives throughout our lives. But we should also live our lives to the fullest as long as we are alive
>
> —*Trude Schei, Consultant Nursing Homes and Health Centers, Bærum Municipality.*

The award-winning Carpe Diem Dementia Village is a groundbreaking project designed to enable individuals with dementia to lead fulfilling lives while remaining connected to their community. The client, Bærum municipality sought to create a nursing home that transcended the institutional atmosphere often associated with such facilities. Their vision was to craft a vibrant, homely environment, tailored to the unique personalities and needs of each resident, with a strong emphasis on basic human values.

C. H. Anglero (✉)
Architect, Partner and Head of Healthcare design, Nordic Office of Architecture, Oslo, Norway
e-mail: chan@nordicarch.com

S. Camplone and C. Porfirione (eds.), *Design for All and the Future of the Planet*, Design For Inclusion 6, https://doi.org/10.1007/978-3-032-21769-1_12

Carpe Diem Dementia Village has set a new standard in healthcare architecture by breaking down the barriers between residents and the surrounding community. The village features 136 communal housing units, 22 high-care dementia units, and shared spaces such as a bistro, urban square, and pathways, all open to the public. The design fosters a sense of community and belonging, allowing residents an active and secure everyday life.

The village is designed with a domestic, small-town atmosphere in mind, featuring varied building heights, roof typologies, and spaces defined by gardens and squares. Recognizing the challenges faced by those with dementia, the design incorporates easily identifiable markers and elements to facilitate wayfinding and create a sense of home.

A key aspect of the design is the distinction between urban and rural environments, reflected in the architectural style and choice of materials. The facades are primarily brick, with untreated wood cladding adding warmth and variety.

The use of durable, maintenance-free materials throughout the facility meets the stringent environmental requirements of the Nordic Ecolabel.

Close collaboration with Bærum municipality, and within the project team, was integral to the project's success. Through workshops and user surveys, the design team gained deep insights into the needs and preferences of the village's future residents. By integrating service design principles with architecture, the project successfully challenged traditional norms for dementia care, resulting in innovative solutions that emphasize what residents can achieve rather than imposing limitations.

Carpe Diem Dementia Village represents a new typology in Norwegian architecture, serving as a model for future healthcare facilities. It has become a reference project not only for retirement homes but also for other sectors exploring inclusive design.

Camilla Heier Anglero*, Architect, Partner at Nordic Office of Architecture, and Head of Healthcare Design will present the architectural and ecosystem perspectives of Carpe Diem Dementia Village. The presentation will highlight the project's approach, value, and its role as a national and international model for inclusive design. Camilla will also discuss the importance of the project owner's innovative brief, the willingness to challenge academic norms, and the trust placed in architects and landscape architects to realize a transformative vision.*

Client: Bærum Municipality.
Location: Bærum, Norway.
Completion date: December 2017–2020.
Total building area: 18.000m^2.
Construction cost: 735 mill NOK.
Architect: Nordic Office of Architecture.
Landscape Architects: Bjørbekk & Lindheim.
Interior designer: Nordic Office of Architecture/Cadi.
Buildingsystems: Norconsult.
General contractor: HENT.

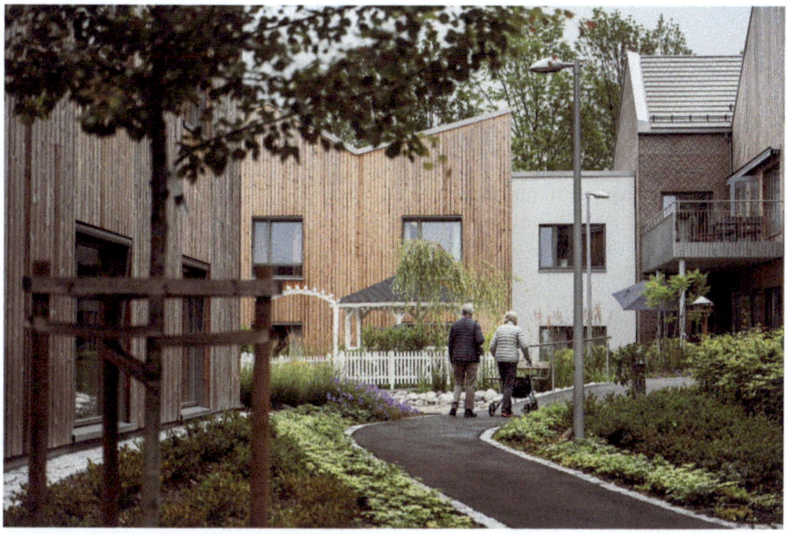

Photo: Nordic Office of Architecture / Benjamin A. Ward

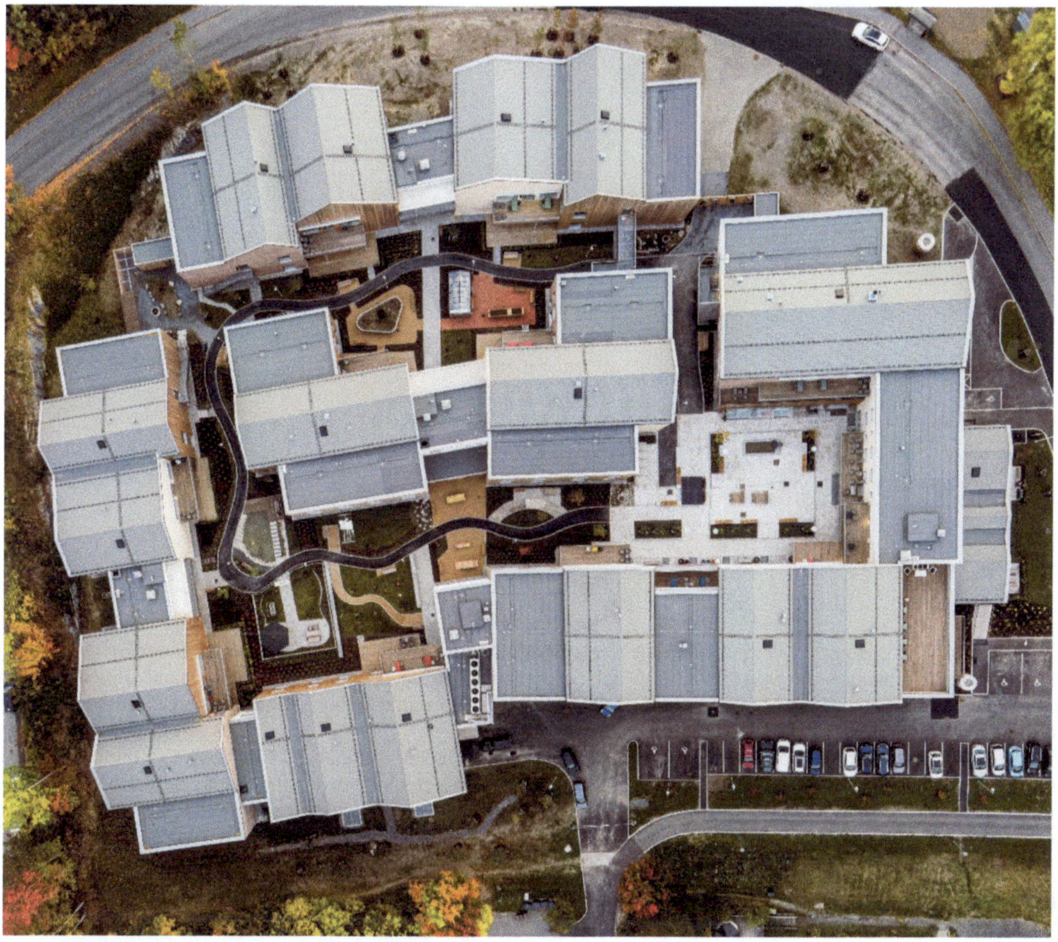

Photo: Nordic Office of Architecture / Benjamin A. Ward

References

Links

DOGA. Hovedvinnere.av.DOGAs.Innovasjonspris.for.inkluderende.design|DOGAGjeve.priser.til.inkluderende.arkitektur|DOGA

Nordic Office of Architecture. Carpe Diem Dementia Village–Nordic Office of Architecture (nordicarch.com).

Possibilities of Adaptation of Seniors Apartments in Existing Residential Concrete Panel Buildings Consistent with Inclusion

L'ubica Selcová, Natália Bošková Filová, and Martina Bezáková

Abstract

The environment has a significant impact on people, their health and wellbeing. Regarding older people, but also children and people with disabilities, the impact of the environment is even more accentuated due to the increased need for accessibility, clarity and coziness of spaces. This academic paper presents possibilities of floor adaptation of purpose-designed prefabricated apartment buildings—homes for older people—built in the ZTB (Unified Type Bratislava) construction system. Large numbers of comparable buildings have been constructed in Slovakia, Europe and worldwide in the second half of the twentieth century, and therefore the selected case study has international application to a large extent. The process of deinstitutionalization of these large-capacity facilities is focused on creating an inclusive and universally accessible, flexible environment providing mixed forms of housing—from flats for seniors to rental apartments for different age groups. Researched designed adaptation solutions reflect demanding structural-static requirements of the panel house; heating, ventilation and air conditioning; fire safety; accessibility and overall principles of universal design. In addition to all these aspects, aesthetic and psychological factors of the environment and biophilic design are considered in recommendations for adaptation. The interior design is therefore proposed in harmonious colors and natural materials with the aim of a friendly and homely feeling.

Keywords

Adaptation · Prefabricated apartment building · Universal design · Social services · Deinstitutionalization

1 Introduction

The need for adaptation and revitalization of senior apartments is currently touching a large number of existing residential concrete panel buildings in Slovakia. One of them is a purpose-designed prefabricated building with a large-capacity facility and flats for seniors. Comparable buildings are widespread throughout Europe and beyond, and mostly serve the residential function.

L. Selcová · N. Bošková Filová (✉) · M. Bezáková
Faculty of Architecture and Design Slovak University of Technology in Bratislava, CEDA–Centre of Design for All, Bratislava, Slovakia
e-mail: natalia.filova@stuba.sk

L. Selcová
e-mail: lubica.selcova@stuba.sk

M. Bezáková
e-mail: martina.bezakova@stuba.sk

S. Camplone and C. Porfirione (eds.), *Design for All and the Future of the Planet*, Design For Inclusion 6, https://doi.org/10.1007/978-3-032-21769-1_13

The research aligns with a wider national project in Slovakia "Deinstitutionalization of Social Service Facilities—Support for Transformation Teams" realized from 2018 to 2023 involving predominantly large-capacity social service facilities. The aim of deinstitutionalization (DI) is a transition from institutional care to community services. The DI process is in this case study intended to ensure an inclusive and universally accessible, flexible and welcoming environment providing mixed forms of housing—from apartments for seniors to rental apartments for different age groups, where mixed accommodation creates a more heterogeneous community.

The House of the Third Age (HTA) facility for older people in Bratislava was one of the involved facilities in the DI project and serves as a case study for this research. The vision of the transformation is to improve clients' quality of life. During the DI project, a Transformation Plan was developed, which declares the necessary changes in several aspects including the physical environment. Research and design team composed of the authors of this paper explored architectural options for adaptation of housing for older people in the facility under study.

Thematic spheres of the research presented in this paper are composed of: (1) Human factor; (2) Spatial solutions; (3) Construction; (4) Heating, ventilation, and air conditioning (HVAC); (5) Fire safety; and (6) Interior design with each of these areas being addressed theoretically and in the recommended design of solutions.

2 Methods

After analyzing the situation and theoretical desk research, in situ field research was realized in the HTA facility. In cooperation with employees of the facility, wishes of the current residents and employees were examined in a survey using questionnaires with the aim of discovering a basis for designing solutions appropriate to people's needs. To detect spatial barriers of the building, an accessibility audit sheet developed by CEDA–Centre of Design for All at the Faculty of Architecture and Design STU in Bratislava was used (Čerešňová & Fratričová, 2023). Then, possible adaptations of the apartment building were tested using research-by-design. Researched premises include (1) social community spaces; (2) flats for seniors and rental flats for a wider target group; (3) bathrooms of flats; and (4) premises for information and nurses. All proposed spaces meet the requirements of accessible use, some of them are even designed in a way that maneuvering of a person lying on a bed is possible, i.e. a maneuvering diameter of not only 1500 mm, but in many areas 2100 mm is planned.

3 Research Findings

3.1 Human Factor

Historical Context. In the second half of the twentieth century, the urban population grew rapidly causing changes in the construction of residential buildings. Large-scale social service facilities were being built for people who needed support and care. In larger cities, they were prefabricated buildings, which were not universally ('barrier-free') accessible. In smaller towns, former mansions were often adapted for this purpose, which were not suitable for it. Today, the need for more than just physical accessibility to social services is increasingly coming to the fore.

In Slovakia, several prefabricated systems were used for social service facilities. The HTA, built in 1985 as a purpose-built facility for older people located in Bratislava, is a representative example for the case study.

Actual State of the Facility. The prefabricated 12 floor building contains a large-capacity facility and flats for seniors. It is a home to 263 seniors with varying degrees of dependency up to 24/7 care, who are provided with social services and live in one-or two-room apartments. In total it has 14 two-room flats with 2 clients, 143 one-room flats with 1 client, 49 day-care rooms with 2 clients. On each floor are symmet-

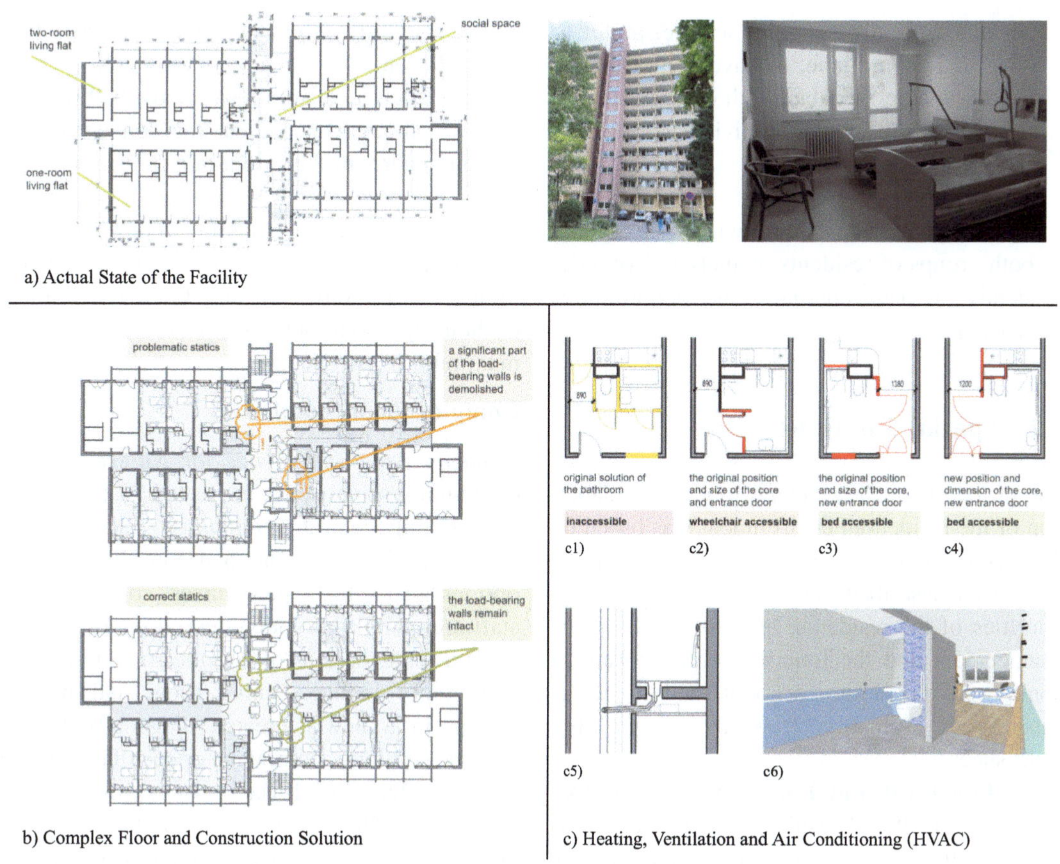

Fig. 1 Set of images: **a** Actual State of the Facility; **b** Complex Floor and Construction Solution; **c** Heating, Ventilation and Air Conditioning (HVAC)

rically placed 18 one-room apartments and 2 two-room apartments, both with bathrooms. The interior of the building is currently not fully barrier-free accessible (Fig. 1a).

Survey Research Findings. Social service facilities for seniors according to the Act on Social Services No.448/2008 Coll. shall be provided to a person who has reached retirement age and is dependent on the assistance of another person and whose degree of dependence is at least IV, or to a person who has reached retirement age and needs the service in such a facility for other serious reasons. According to the survey, seniors in the HTA give the following reasons why they are or will be recipients of the service with the percentages indicated: (1) I will need help in the near future (27,5%); (2) I need social supervision because of cognitive disorders (26%); (3) I need nursing help (24%); (4) I need affordable housing (22,5%); (5) Husband or wife needs help today (4,8%); (6) I need accessible housing (2,2%).

As part of the survey, research of the wishes of the residents and employees was conducted also using questionnaires. Desires and wishes of the inhabitants were divided into two groups:

(1) The clients with lower degree of dependence prefer to live in their apartment unit as long as possible, to stay independent, to have contact with other people, to find close friendships with other recipients, to have a choice of more activities to fill their free time.
(2) The clients in the all-day care department with higher degree of dependence, especially

in this type of residential service, prefer to live in a room alone, to have help available (staff, doctors), contact with other people, to have a choice of more activities to fill their free time.

It can be stated that some wishes are shared by both groups of residents, namely to live independently, to have company and opportunities for activities.

3.2 Spatial Solutions

The proposal for the research-by-design adaptation of the residential environment was based on the results of the survey of the clients' wishes and needs described above. Specific ideas and priorities of the residents, but also needs of the management and the founder of the retirement home were taken into account, and in accordance with evaluation from the accessibility audit sheet.

Architectural and Functional-Operational Solution of the Building's Typical Floor. Based on the floor plan solution of the building for adaptation were selected one-room apartments, which are located on the floor most and originally a communication space serving social and community function. The layout, operational and functional problems were identified and addressed. In the flats, these are mainly the barrier entrance areas and bathrooms of the flats, poor and impersonal furnishings of the flats, the communication and community space not properly lighted and ventilated, the absence of information points and spaces for nurses, inadequate spaces for storage of utilities and others.

The applied research carried out in connection with the architectural design of the adaptation was a comparison of:

(1) two alternative options for the extension and relighting of the common rooms in the central part of the floor layout;
(2) two alternative options for the spatial design of a one-room apartment with a high degree of variability, together with a barrier-free bathroom;
(3) two alternative solutions for the bathroom;
(4) two alternative options for converting the one-bedroom flat into a nurses' room.

The large degree of variability represents, with the same layout with minor variation, the use of the flat for day care and for a rental form of housing, both options for occupation by a maximum of two people.

Complex Floor Solution

Common Area with Space for Nurses. A new community social space is created by expanding the existing central community space by two rooms (replacing 2 one-bedroom flats). Both options also offer a new nurses' room space–a staff area with a reception desk (Fig. **1**b).

The first option offers the function of a community lounge, dining room and nurse's station in a newly created diagonally better-lit and ventilated communal space. The second variant offers the function of a nurses' and community dining room with a larger dining table and a smaller lounge with cross ventilation, albeit a little less well lit. Both options allow for movement and maneuverability of a person in a wheelchair and a person in a reclining bed.

One-Room Apartment for Day Care Unit. A variant solution has been created with a single bed or a double bed for partners, with a fully accessible bathroom with double sliding or opening doors for showering on a shower lounger (Fig. 2a). The option offers the transfer of a client on a reclining bed down the corridor from the room directly to the common area. The kitchen is designed with a simple flexible worktop with a sink and a kettle. Other equipment can be added according to the client's needs, wishes and state of health.

One-Room Rental Apartment. There are proposed variant solutions with a single bed or with a double bed for partners, both with an accessible bathroom and kitchenette (Fig. 2b). For the rental form of housing, it is possible to use both variants of bathrooms shown on the following floor plans: (1) L-shaped

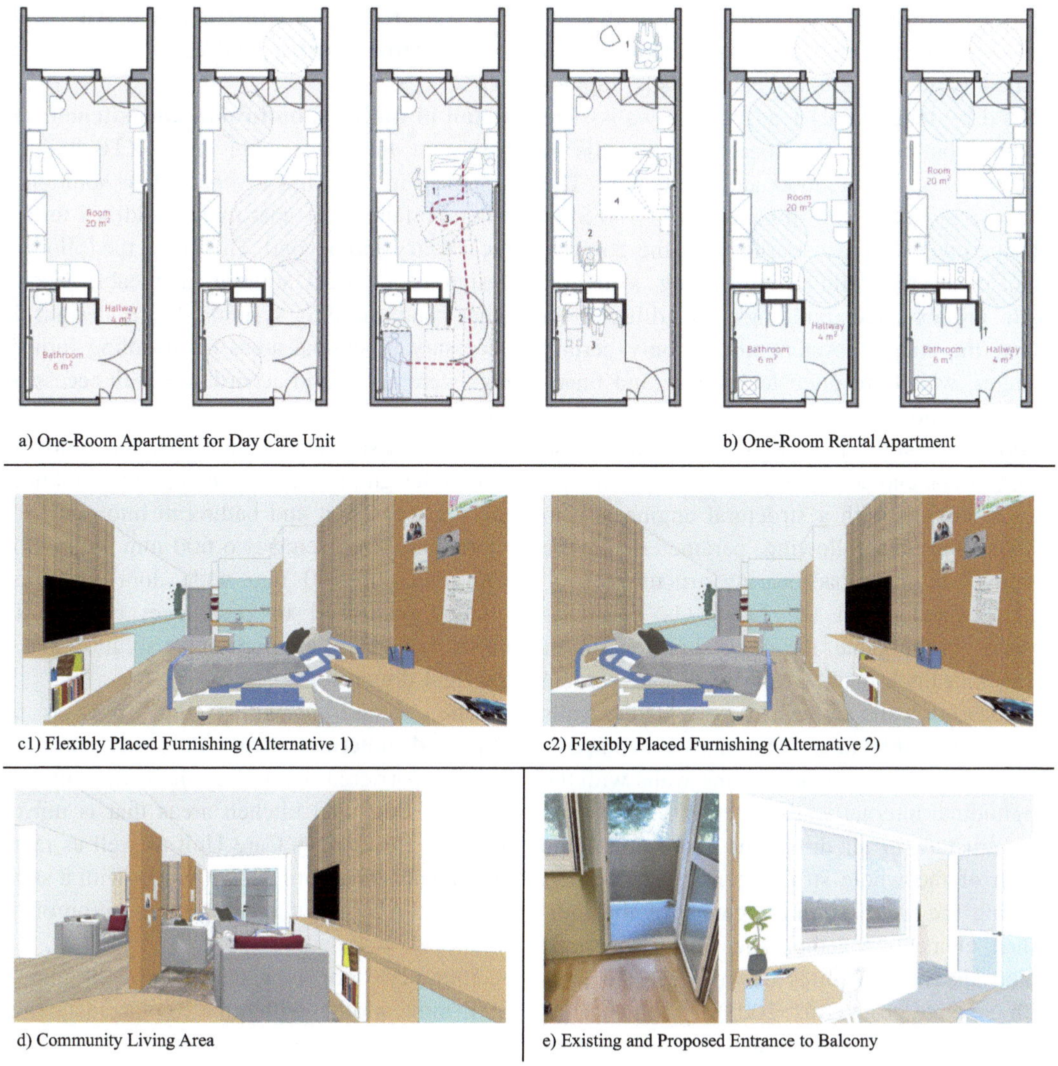

Fig. 2 Set of images: **a** One-Room Apartment for Day Care Unit; **b** One-Room Rental Apartment; **c1** Flexibly Placed Furnishing (Alt. 1); **c2** Flexibly Placed Furnishing (Alt. 2); **d** Community Living Area; **e** Existing and Proposed Entrance to Balcony

floor plan with a narrower corridor with a width of 890 mm, (2) rectangular floor plan with a wide corridor of 1380 mm and with the possibility of using one or both door wings of the entrance door to the bathroom to open. For renting especially to the older people, to allow age-in-place, alternative (2) is preferred, which also allows for bed showering although it is financially and technically more demanding.

3.3 Construction

Since 1972, row and slab apartment buildings in the ZTB (Unified Type Bratislava) construction system have been built in Slovakia. The ZTB system is characterized by a system of transverse walls, two communication cores, and communication spaces in the middle, in this case used as a social space. The debarriered facility for seniors is designed as a slab apartment building with minor modifications of the typified solution. In slab buildings, the support system is formed by transverse load-bearing walls, supplemented by longitudinal load-bearing walls. The modular outline of the load-bearing walls is formed by spans of 3600 and 2700 mm. The staircase module has a dimension of 2700 mm. All apartments are accessible from the corridor area, connected with a protruding staircase. Such prefabricated buildings are difficult to adapt, the support system allows only certain changes, which are often technically and financially demanding.

For the design of the debarrierization of the living, social and accessory spaces, it was necessary to consult with a structural engineer, who pointed out the following parameters of the existing building's load-bearing structures:

The envelope in the façade is non-load-bearing. In the gable, the outer skin consists of a 150 mm thick load-bearing wall. The vertical load-bearing walls are reinforced concrete walls of 140 mm thickness (150 mm in composition). The transverse load-bearing walls with the longitudinal internal walls together form a rigid box structure in all directions. The monolithic action of the whole structure was achieved by welding the contact reinforcement of the individual elements together and filling the horizontal and vertical joints between the elements with concrete grout (Búš, 2008).

In the design, simple and almost unproblematic interventions to non-load-bearing structures were selected by the structural engineer: cutting a door opening in a partition, removing a partition, replacing the housing core, which may, however, change the vertical load on the load-bearing structures. Interventions in load-bearing structures should be minimized and recalculated by a structural engineer at a later stage. For this reason, the design was finalized in the variant with minimal interventions in the load-bearing structures (Fig. 1b).

3.4 Heating, Ventilation and Air Conditioning (HVAC)

Detail of hallway, bathroom and kitchen. The bathroom and toilet cabin in the ZTB building system (Fig. 1c1) has a very similar solution to other prefabricated apartment buildings around the country and beyond. Therefore, the following design is a model solution applicable also to other panel building systems. In the design, the necessary functional areas for handling furnishings were applied in accordance with accessible use.

In the first design (Fig. 1c2), the following spatial design practices were applied: combining the separate toilet and bathroom into one bathroom space; replacing two 600 mm wide doors with one 800–900 mm wide door; bathroom fittings positioned to best meet accessibility needs, but this is a compromise solution due to the tight dimensions.

The second proposal (Fig. 1c3) is an improved alternative to the first proposal of a debarriered bathroom, hallway, corridor, entrance door and kitchen areas that is universally usable for Day Care Unit as well as rental apartments. The third design works with a slight shift of the core, but keeping the position of the door (Fig. 1c4).

Entrance door is extended to 900 mm + 300 mm width, with this replacement also possible from the structural point of view. The corridor to the bathroom and room with a width of 1380 mm allows the movement of a reclining bed or the transfer of a client on a shower bed into the bathroom within the flat.

If the space under the bathtub is slightly drained (e.g. 2.0%) to a drain point in the floor, where the shower floor drain will be installed, it would also be an interesting solution to leave the bathtub. If sufficient space is available in the bathroom with the bath, a barrier-free shower can be retrofitted and the bathtub can be retained (Peráčková, 2022).

The proposed bathroom has a one-level design of the shower and is located on the same floor height as the surrounding areas. This is possible thanks to a floor drain below the ceiling level of the bathroom one floor below. It is covered using a lowered ceiling, so the ceiling above the washbasin is lower than in the other areas (Fig. 1c6). Schematic section (Fig. 1c5) shows the method of fitting and draining a floor drain or drainage grille for a tankless shower in the space below the ceiling using soffit also to accommodate a shower vent. This detail provides a better solution than having to raise the floor level in the bathroom itself relative to the floor of the surrounding spaces due to the floor drain being connected to the existing plumbing core. It can be noted that this solution is ideal for a building with a single owner of all apartments, it would be complicated in an apartment building with different owners.

3.5 Fire Safety

As a large-capacity facility, the Third Age Facility is awaiting, in addition to debarrierization and renovation, also the improvement to fire safety in the form of additional stairways on the gable walls of the building and a new evacuation lift. From the research by Hung et al. (2022), fire cases in Taiwan's old welfare institutions in recent years can be introduced: the locations and reasons of the fire cases have been described as man-made arson in storage room (13 dead and 59 injured); electrical appliances in ward on the eighth floor (6 dead and 28 injured); candles ignited flammable materials on second floor in a flat housing near the stairs (4 dead and 13 injured) etc. There have also been tragic fires of similar facilities in Slovakia, so this is an extremely important aspect.

3.6 Interior Design

As part of the research-by-design, it is important to address both the physical, collision-free and accessible solutions and the psychological impacts of the solution, including humanization in healthcare contexts (Bonnes et al., 2009). The aim is to create welcoming and personal spaces offering flexibility and individual refinement. The design is based on color and material surfaces considering the effect on emotions. Based on biophilic design, wooden surfaces, natural colors and greenery are implemented in the interior. The positive impact of wood on mood in health service environments has been demonstrated by research by Kotradyova et al. (2019).

The one-room apartment is universally furnished with the possibility of placing the bed on both sides of the room according to clients' wishes. To achieve the versatility, the adjacent wall on both sides of the room is fitted with natural wood paneling. In alternative 1 (Fig. 2c1), the bed is isolated from the view of incoming people but the solution gives an airier impression because of the free passage line between the front door and the balcony. In alternative 2 (Fig. **2c2**), the bed allows a view from the bed not only of the exterior, but also of the front door. Easy relocatability of the bed and TV table creates the flexibility. The cork paneling above the desk used to attach pictures is another natural and warm material.

The community living area (Fig. 2d), is designed in a pastel color scheme with interchangeable interior accessories. Parts of the walls are also covered with cork for the aesthetic presentation of the residents' artwork and announcements. The interior is reminiscent of a living room in a family setting. The spacious layout with sightlines to adjacent areas are suitable for navigation for people with various disabilities. There are also more intimate semi-enclosed zones for individual relaxation and activities.

The design sought to promote physical and visual communication with the outdoors. Each room and lounge area are connected to the balcony via an interior ramp. Currently, access to the balcony is a barrier due to a 80 mm step. In the proposal, the balcony floor is raised to the height of the balcony door threshold using an exterior terrace located on tertiles (Fig. 2e). Due to this elevation and for safety, a bar railing has been added, but subtle enough not to block the view.

4 Discussion and Conclusion

In this research, possibilities of adaptation of a specific ZTB panel building in Bratislava were investigated. Perspective successive research is to prepare a methodology for the adaptation of various standardized housing systems and the design of variant forms of individualized housing. It can be intended for seniors and other target groups, or also the general public, to increase the quality of housing and social services. Adaptable space will respect the need for privacy, the need for personalization and flexibility in the use of space, and for socialization—inclusion of people with disabilities in the community. Further follow-up research will focus more on psychological impacts of the built environment and the factors which can positively influence them.

In the researched building presented in this paper, adaptation of community spaces and apartments with the possibility of their universal accessibility and better possibility of providing a high level of care were examined. The research-by-design results propose solutions customized to specific residents' needs, promoting inclusion, diversity and mixing of different target groups of residents.

Given the mass production of prefabricated houses, albeit with some variations, it can be assumed that the research results could be partially implemented in different prefabricated houses serving as social service facilities in various countries.

Acknowledgement The paper is published within the framework of 2 projects:

(1) The national project "Deinstitutionalization of Social Service Facilities—Support for Transformation Teams", which is implemented thanks to the support from the European Social Fund and the European Regional Development Fund under the Human Resources Operational Programme;

(2) The project "Parameters of architectural environment supporting educational effectiveness" funded by the EU NextGenerationEU through the Recovery and Resilience Plan for Slovakia under the project No. 09I03-03-V04-00328.

References

Bonnes, M., Bonaiuto, M., Fornara, F., & Bilotta, E. (2009). Environmental psychology and architecture for health care design. In R. D. Nord (Ed.), *Proceedings of 28th international public health seminar* (pp. 35–41).

Búš, V. (Ed.) (2008). Zásahy do nosných konštrukcií panelových bytových domov (2nd ed.) VVÚPS-NOVA, výskumno-vývojový ústav pozemných stavieb. ISBN 978–80–89073-14-6.

Čerešňová, Z. & Fratričová, S. (2023). *Hodnotenie bezbariérovej prístupnosti budovy* (CEDA).

Hung, S.-M., Wang, S.-C., Chien, S.-W., Su, C.-H., & Chen, L.-P. (2022). The enhancement of fire safety in small-scale senior citizen welfare institutions based on fire protection defense-in-depth strategy. *Sustainability, 14*(6), 3196. https://doi.org/10.3390/su14063196

Kotradyova, V., Vavrinsky, E., Kalinakova, B., Petro, D., Jansakova, K., Boles, M., & Svobodova, H. (2019). Wood and its impact on humans and environment quality in health care facilities. *International Journal of Environmental Research and Public Health, 16*(18), 3496. https://doi.org/10.3390/ijerph16183496

Peráčková, J. (2022). Univerzálne navrhovanie–bezbariérovosť a prístupnosť pre všetkých, p.52, *TZB Haustechnik,* č.5/2022, 30. ISSN 1210-356X.

The Universal Design: Undivided Part of Architectural Expression

L'ubica Selcová and Stanislav Majcher

Abstract

Equal conditions for all, regardless of age, ability, or disability. Universal design puts human diversity at the centre of the design process so that buildings and environments can be designed to meet the needs of all users. It therefore applies to all people regardless of their age or height, as well as to those with any physical, sensory, mental, or intellectual disability. The accessibility of buildings is one of the important criteria determining the quality of use and is an essential condition for the sustainable development of the residential environment. Barrier-free access should currently be a self-evident solution for every publicly accessible building. We can achieve it in several ways. It is ideal to design the building from the beginning of the project so that the entrance area is at the same height level as the access road. Achieving barrier-free access is an intervention in the design of the building, and many times it results in the form of an additions that look uncomfortable in the whole context of the building. The overall aesthetic expression of the building and its comprehensive accessibility for everyone should be the two main goals of every designer's work (Samová et al., 2008).

Keywords

Universal design · Accessibility · Ramp · Reception · Window · Bathroom

1 Introduction

"Whether you are eight or eighty years old, you want your world to work for you." (Exhibition "Unlimited by Design" at the Cooper Hewitt, Smithsonian Design Museum in New York, 98/99). This quote presents the basic principle of universal design. It is also the key motto of the architectural studio's ARCHICO/Jablonský, Majcher work.

In the design of the built environment, it is necessary that people can access it, enjoy it to the greatest extent possible, as independently and naturally as possible, without the need for further modifications or specialized solutions.

2 Methods

After analyzing the situation and theoretical desk research, in situ field research was realized in different locations of the built environment described in this practical contribution.

L. Selcová (✉) · S. Majcher
Faculty of Architecture and Design, Slovak University of Technology in Bratislava, CEDA—Centre of Design for All, Bratislava, Slovakia
e-mail: lubica.selcova@stuba.sk

S. Majcher
e-mail: stanislav.majcher@stuba.sk

S. Camplone and C. Porfirione (eds.), *Design for All and the Future of the Planet*, Design For Inclusion 6, https://doi.org/10.1007/978-3-032-21769-1_14

Any process of designing the built environment should be preceded by desk research which, after taking into account input parameters such as the context with the environment, the brief, the legislation, the author's handwriting, leads to the verification of the initial ideas by research by design. The concept as a materialized idea is an experiment which is further validated by the following design process or by research during the design process.

Over the past decade, there has been a particularly lively debate in architecture and allied fields about the extent to which "design" is or should be a template, or more broadly perhaps, a new "paradigm" for research in creative or professional domains. It is possible to deal with several facets of the relationship between design and research (Groat & Wang, 2013).

3 Research Findings

The subject of research, and the main essence of the architectural work of the studio Archico, are buildings providing social and health services. Especially these typological types of buildings must provide their visitors and clients with a universally accessible environment without barriers.

According to a source from the Statistical Office of the Slovak Republic, an accessible environment is a necessity for 10–30% of people, useful aid for 50% of people, and a greater comfort for 100% of people. Accessibility of the environment can be evaluated on the basis of a number of well-designed building parameters: barrier-free entrance and communication spaces, barrier-free visual contact with the outdoors, and barrier-free sanitary facilities.

3.1 Barrier-free Entry—A Ramp as a Part of the Architectural Expression

The accessibility of buildings is one of the important criteria determining the quality of the built environment and is an essential condition for the sustainable development of the buildings for social services and healthcare. "The European definition of sustainable intelligent building includes architectural criteria focused on the quality of the interior and the quality of the design. These criteria confirm the relevance of the architectonic concept in the creation of sustainable intelligent buildings in Europe" (Puškár et al., 2019).

Barrier-free access should currently be a self-evident solution for every publicly accessible building. We can achieve it in several ways. It is ideal to design the building from the beginning of the project so that the entrance area is at the same height level as the access road. If this is not possible, then a ramp comes as a solution, or technical equipment in the form of a stair platform or better solution—elevator. However, this method depends on several factors and does not fully meet the conditions of "the most independent and natural" movement. Therefore, it should only be used in cases where a ramp is not an option.

Achieving barrier-free access is an intervention in the design of the building, and many times it results in the form of an additional sticker that looks awkward in the whole context of the building.

However, when we realize the shape potential that the ramp provides, we can use it for the benefit of architecture. The ramp subsequently becomes an integral part of the architectural expression of the building.

The example of the balneotherapy reconstruction in Lúčky spa, where the composition of the extended terrace from the cafe is completed by a ramp. The railing is made of glass. The handle is attached directly to the safety glass and is 90 cm high. The shortcoming of the solution is the missing handle, which should still be at a height of 75 cm. The guide bar for the wheelchair wheels is omitted in this case, as its function is taken over by a full wall of safety glass. The tread layer is solved in the form of a large-area concrete pavement, which has a natural anti-slip character.

In the case of Dúbravská oáza the ramp is designed not only to overcome the height level,

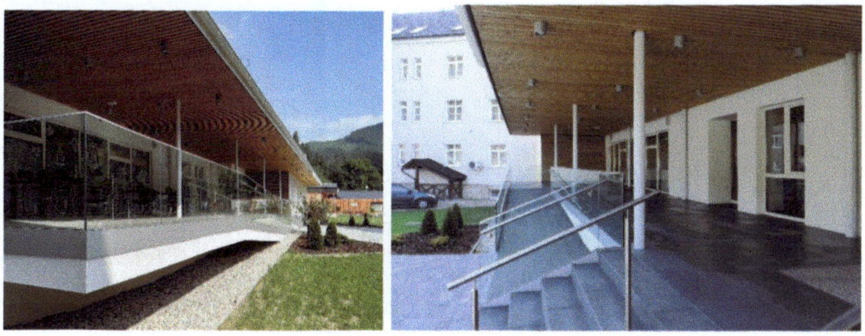

Fig. 1 Set of images: Barrier-free ramp as part of the architectural expression of the building, Lúčky Spa, 2019 (architects: M. Jablonský/S. Majcher, foto: M. Jablonský)

but it is also a “path” that provides a walk around the green garden. The overlay with a reinforced awning subsequently accentuates the main entrance, which is already subconsciously legible from a distance. This glass covering, in combination with anthracite concrete paving, also creates an interesting “greenhouse” effect for the covered space. So, in the winter months, when the snow gets from the sides into the entrance area, it melts very quickly (Fig. 1).

3.2 Entrance Spaces

The entrance to the building must be well perceived with strong architectural contrast used. In the case of a vestibule, the depth requirements are different if it is a sliding door and different if it is an opening door. With automatic sliding doors, the required depth of the vestibule is min. 160 cm. For opening doors, the required depth is 210 cm, which, however, cannot always be achieved due to space requirements. In picture no. 4 is a vestibule solution in cramped conditions, where the handling space for a wheelchair user is extended to one side, thus eliminating the depth to 170 cm. In all cases, if it is a sunlit door, it must be equipped with a signal strip at eye level (160 cm). The glass used on the entrance door must be safety glass (VGS—class 2B2, on both sides of the door). The clear width of the entrance door should be min. 110 cm (ideal if the door is equipped with an additional wing).

Reception and information desks should be accessible and understandable for everyone. This includes all visitors as well as services working on the counter side. They should be placed prominently and in a direct position in relation to the main entrance to the building to allow for easy identification. At the same time, away from potential sources of noise.

The work surface should be designed at two different levels to facilitate use by people of different heights (sitting or standing). The lower working surface should be a maximum of 76 cm above the floor level. If it is necessary to sign documents, it should have free space for legs up to a depth of 70 cm. The upper work surface should be at a height of 95–115 cm above the floor level. At the same time, the height of 115 cm provides sufficient space for placing the monitor under the top plate. There should be sufficient handling space on both sides of the reception or service desk (min. 244 × 244 cm in front of the desk). The edge of the counter should visually contrast with the work surface so that it is easily identifiable. If receptions and service desks are required to be glazed for security or other reasons, they should be clean and free of plastered obstructions.

The placement of artificial lighting should be carefully considered in relation to any glazed surface to avoid the possibility of glare and reflection as this can impair visibility and communication. Where counters are glazed, a public address system (also called a voice transmission system) should also be considered. This is likely to benefit everyone, including the hearing impaired, as the clarity and volume of speech is often reduced by the presence of glazing (Figs. 2 and 3).

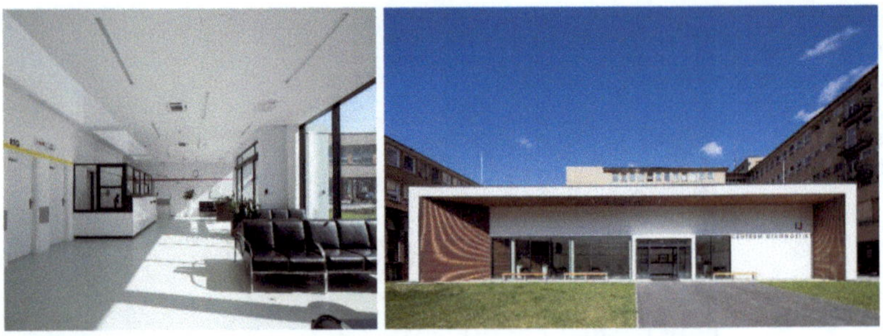

Fig. 2 Barrier free entrance and variable height of the reception desk and contrasting solution of the edge of the desk with the worktop, Center for diagnostics—new pavilion, Hospital Považská Bystrica, 2022 (architects: M. Jablonský/S. Majcher, foto: T. Manina)

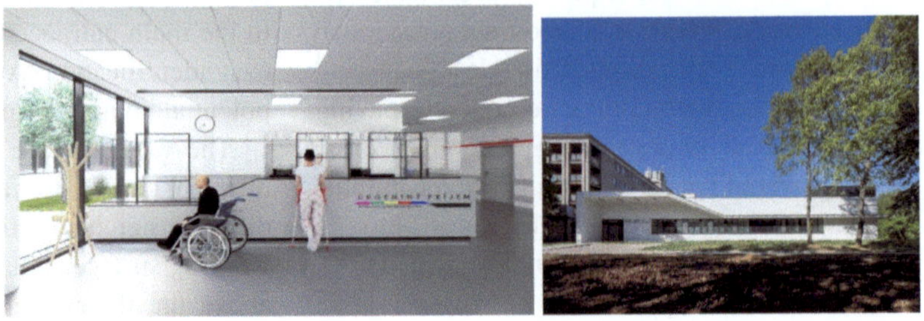

Fig. 3 Barrier free entrance and variable height of the reception desk and contrasting solution of the edge of the desk with the worktop, Emergency—new pavilion, Hospital Bojnice, 2022 (architects M. Jablonský/S. Majcher, foto: T. Manina)

3.3 Window—A View for all

The windows are essential in buildings to provide natural lighting and ventilation. They also provide a visual link to the outside environment, which can help with orientation and wayfinding within a building or site. A good view through a window can greatly improve the usability and enjoyment of a room or space in a building. The location, size and number of windows depends on many factors, including the requirement for natural lighting and ventilation. In rooms where it is important for people to see well through the window, the position of the sill and all horizontal elements of the window frame should be carefully considered.

For people sitting to enjoy an unobstructed view, the window area between 90 cm and 120 cm above floor level should not contain any dividing elements. If the windows are designed to be opened manually, the controls should be accessible. Ideally, they should be placed between 80 and 110 cm above the floor level. All window handles should be able to be operated with one hand, preferably with a lever. The use of electrically operated window openers should be considered in all cases where handles are likely to be out of reach. Such devices can be controlled by a wall switch or a hand-held remote control. If wall switches are provided, they should be clearly marked, visually contrasting with the surrounding surfaces. Their location should be at a height of 75–120 cm above the floor level.

If the height is lower than 50 cm, then the parapet is considered a stepping surface, and it is necessary to ensure that 90–120 cm above it (depending on the height above the ground) there is no possibility of falling. This entire height must be protected by safety glass. If the windowsill is over 50 cm, then the anti-overflow protection only needs to be done at the height, which is calculated from the floor, which makes it possible to partially save funds on the surface of the safety glass (Fig. 4).

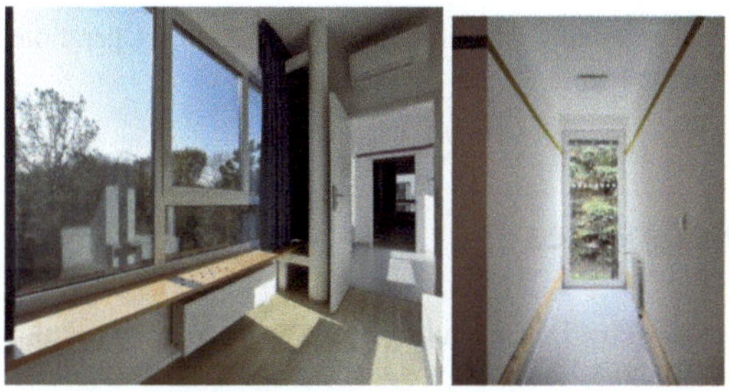

Fig. 4 Presents a solution with a window with a lowered parapet, or without the parapet in reconstructed and extended Social Services Home Nosice—Púchov, 2022 (architects M. Jablonský/S. Majcher, foto: S.Majcher)

3.4 The Bathroom

Bathrooms are a place of daily need. Their universality determines the sustainability of housing at any age or life situation. It is also the place where most injuries generally occur due to falls in combination with poorly designed furniture and slippery floor conditions. Universal design should already be the basis for the design of the bathroom, although it does not have to be equipped with handrails and other accessibility aids. However, if necessary, its spaciousness will allow it to accommodate wheelchairs, crutches, or handrails in the shower and near the toilet. The additional reconstruction of inappropriately oversized premises is consequently financially demanding and is the result of compromises.

In the case of the floor, it is important to think about safety. Falls are the leading cause of death for the elderly, although slips are dangerous for anyone of any age. For this reason, the optimal solution is a combination of non-slip flooring with underfloor heating, which speeds up drying. When tiling, it is important to think about its height, which can be designed up to the ceiling or, from an aesthetic point of view, it should catch the edges, such as the upper edge of the cladding frame.

A spacious shower corner without a tub is the most optimal solution for the bathroom in terms of universality of use. It should be equipped with a folding seat with handles on the sides. The height of the seat should be 50 cm above the floor. Shorter height can put extra pressure on the knees when forcing into a squat position. The height of the control lever battery should be from 85 to 100 cm, so that a person does not have to strain even in the case of a mild disability. The solution of a shower without a tray requires that the drain be handled by a drain, which needs a certain height of the floor (min. height of the drain is 6 cm + the slope of the floor). This detail is an essential condition during renovations, when it is possible to implement a shower without a tub. Otherwise, it is necessary to raise the floor, which will generate an increase in door openings, or attacks.

The toilet should have a seat at a height of 50 cm and should be equipped with one fixed and one folding handle. While on one side there should be a handling space for a wheelchair with a width of min. 80 cm and on the other hand it is important to observe the dimension of min. 27 cm from the wall.

The mirror is a specific part of the bathroom, it enlarges and illuminates the space. From an architectural point of view, we are often prejudiced against folding mirrors (price, small size, "extra" folding mechanism). In the case of using a glued mirror, its optimal location is one that also takes into account the sitting person. In this case, the lower edge of the mirror should be at the level of the upper edge of the sink.

4 Discussion and Conclusion

A characteristic feature of the population is its diversity. Life expectancy is increasing, and we expect more and more to maintain our usual activities in public as well as the private sphere as we age. However, many human capabilities decline with age, such as mobility, dexterity, stamina, strength, hearing, vision or memory, which fundamentally affects the sustainability of the use of living space. For this reason, universal design in architecture cannot be just something extra, but must be an integral part of its DNA, from the initial concept, including the perception of the interior space. Older people often face health problems, including various eye conditions, which need to be considered when planning accessible and inclusive environments. Adapting interior spaces is a critical factor in supporting the needs of older adults, and promoting their independence, safety, and comfort (Kotradyova et al., 2023).

In the realized buildings of the studio ARCHICO presented in this paper, were explored the possibilities of extensions, revitalizations or conversions of existing spaces: Reconstruction and extension of Social Services Home Nosice—Púchov, The Emergency Pavilion, Hospital Bojnice, Center for diagnostics—new pavilion, Hospital Považská Bystrica, Reconstruction and extension of Elderly and Alzheimer Center, Bratislava, and the reconstruction and extension of Balneotherapy, Lúčky Spa.

The proposed facilities also demonstrate through their active operation their universal accessibility and thus better ability to provide a high level of care. The results of the research propose an adaptation adapted to the specific needs of the residents, promoting inclusion, diversification and mixing of the different target groups of residents.

Acknowledgement The paper is published within the framework of the project:

The national project "Deinstitutionalization of Social Service Facilities—Support for Transformation Teams", which is implemented thanks to the support from the European Social Fund and the European Regional Development Fund under the Human Resources Operational Programme.

References

Barrier-Free Design Guide, Based on the Alberta Building Code 2014. (n.d.). ww.safetycodes.ab.ca

Building for Everyone: A Universal Design Approach Facilities in buildings, Centre for Excellence in Universal Design, National Disability Authority. (n.d.). www.nda.ie

Groat, L., & Wang, D. (2013). *Architectural research methods*. Wiley & Sons, Inc.. https://nexosarquisucr.wordpress.com/wp-content/uploads/2016/03/architecturalresearchmethods-groat_wang.pdf

Kotradyova, V., & Hencova, M. (2023). *Color in the environment for older adults*. ALFA:Architecture Papers of the Faculty of Architecture and Design STU. https://doi.org/10.2478/alfa-2023-0021

Puškár, B., Vráblová, E., Kočlík, D. (2019). The Quality of Indoor Environment of Intelligent Building-Global Phenomena of Sustainable Architecture. WMCAUS, 2019. https://doi.org/10.1088/1757-899X/603/3/032102

Samová, M., Čerešňová, Z., Rollová, L., Korček, P., & Majcher, S. (2008). *Tvorba bezbariérového prostredia: Základné princípy a súvislosti*. Eurostav. ISBN:978-80-89228-10-2.

Good Practices: Products and Services

Inclusive Wayfinding at the Academy of Fine Arts and Design in Katowice

Anna Pohl and Agnieszka Nawrocka

Abstract

The article addresses issues related to the adaptation of signage systems in public spaces in Poland to meet the needs of people with disabilities, in particular those with vision impairment or losswith. It describes the process of implementing accessibility in terms of wayfinding systems, including accessibility audits, concept design, and user testing. Key issues in current systems are discussed, including poor visual consistency with existing designs, lack of user consultations, and poor material quality. The authors highlight the importance of inclusivity in design, emphasizing the role of an interdisciplinary approach. The project involved experts from various fields, including psychology, visual communication design, and inclusivity, enabling the development of comprehensive solutions tailored to user needs. A central element of the process was the development of visual information systems to integrate accessibility features into existing architecture and aesthetics, specifically designed for individuals with visual impairments. Examples include tactile maps, 3D building models, Braille signage, and other tools to support spatial orientation. Prototyping and testing solutions with users—both sighted and visually impaired—constituted a critical part of the design process. This approach enabled adjustments that improved the functionality of the systems. The authors emphasize that designing inclusive information systems for one group with specific needs often enhances the usability of the solution for other users as well. The importance of user research as the foundation for creating effective and practical systems is strongly underscored. Based on their findings, the authors developed universal guidelines and educational materials. These resources are intended to assist other institutions in implementing inclusive solutions, while also establishing standards of best practices for designers and managers of public spaces.

Keywords

Wayfinding systems · Inclusive design · Accessibility · Visual impairment · Research through design

1 Context

In recent years, Poland has introduced funding alongside regulations aimed at adapting public spaces to meet the needs of people with disabil-

A. Pohl (✉) · A. Nawrocka
Faculty of Design, Academy of Fine Arts and Design in Katowice, Katowice, Poland
e-mail: anna.pohl@aspkat.edu.pl

A. Nawrocka
e-mail: agnieszka.nawrocka@aspkat.edu.pl

S. Camplone and C. Porfirione (eds.), *Design for All and the Future of the Planet*, Design For Inclusion 6, https://doi.org/10.1007/978-3-032-21769-1_15

ities. A typical process begins with an accessibility audit, followed by changes in architecture, necessary equipment, and wayfinding systems. Unfortunately, most of the implemented solutions have significant shortcomings, including low visual and material quality as well as inconsistency with existing visual systems or the character of interiors and buildings. Decisions regarding the scope of accessibility projects are often made by officials and institutional administrators based on legal regulations, without verifying the genuine needs of users or consulting designers specializing in inclusive solutions. Frequently, the implemented elements are not reviewed for usability, leading to the replication of flawed solutions across other public institutions.

In Poland, discussions on accessibility for blind and visually impaired users have increasingly emphasized the role of multisensory orientation systems in public space (Kalbarczyk, n.d.). Contemporary wayfinding systems are understood not only as collections of signs, but as integrated spatial communication systems combining architecture, information design, and user cognition (Arthur & Passini, 1992). Effective orientation in space relies on the creation of mental maps and recognizable spatial landmarks (Lynch, 1960). Recent Polish reports concerning the needs of people with disabilities indicate that accessibility barriers in public institutions still significantly affect independent navigation and participation in everyday activities (Państwowy Fundusz Rehabilitacji Osób Niepełnosprawnych: Raport końcowy z badania potrzeb osób niepełnosprawnych w Polsce. 26, n.d.). Many existing accessibility implementations are based primarily on legal compliance rather than user-centered design processes. As a result, accessibility elements are often perceived as visually detached additions rather than integrated components of the architectural environment. Inclusive design approaches emphasize that accessibility solutions should be embedded into the overall spatial and visual identity of public institutions rather than treated as supplementary features (Clarkson et al., 2003; Norman, 2013).

2 Challenge

To address the lack of exemplary cases and guidelines for designing system and inclusive wayfinding solutions in Poland, a project was undertaken to develop best practices through a research-through-design approach. Research through design treats the design process itself as a method of inquiry, combining iterative prototyping, reflection, and user evaluation to generate practical and theoretical knowledge (Zimmerman et al., 2007). To achieve this goal, an interdisciplinary research and design team was established, consisting of experts with the necessary skills to carry out this task. The interdisciplinary character of the team reflected contemporary inclusive design methodologies, which stress the importance of collaboration between designers, researchers, accessibility specialists, and users themselves (Clarkson et al., 2003; Sanders & Stappers, 2008). The team members included psychologist and researcher Patrycja Rudnicka, PhD, inclusivity specialist and designer Gregor Strutz, visual communication designers and researchers Agnieszka Nawrocka, PhD, and Anna Pohl, PhD, and product designer Kinga Pawlik, MA.

The research setting was the campus of the Academy of Fine Arts and Design (AFAD) in Katowice, comprising three architecturally diverse buildings constructed in different decades. The academy has roots in printmaking and design and is known for fostering creativity and respecting heritage. The challenge was to create an inclusive wayfinding system to reflect the unique identity of the institution while accommodating the diversity of its users and their varying needs.

3 Project Objectives

The goal was to integrate accessibility elements into the existing visual information system, ensuring a cohesive and inclusive design that would not differentiate or marginalize any user group. Guided by the principles of inclusive and universal design, the project sought to develop solutions accessible to the widest possible range of users (Kowalski, n.d.; Clarkson et al., 2003). It also aimed to establish guidelines supporting designers in creating legally compliant wayfinding systems, as well as educational materials for public building administrators presenting implementation strategies aligned with current Polish accessibility standards and recommendations (Urząd Miasta Warszawy: Standardy zapewniania dostępności, n.d.).

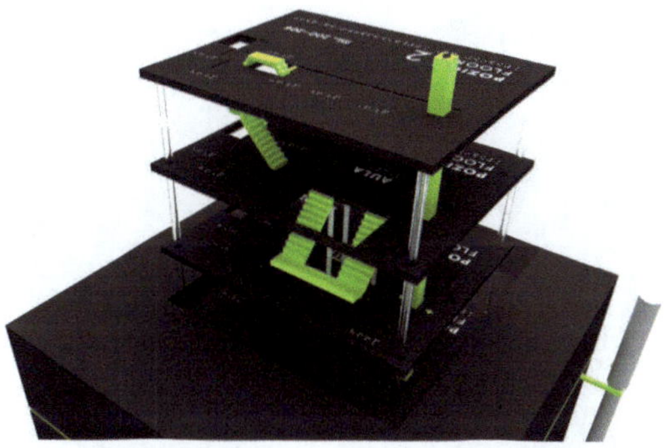

Fig. 1 Spatial model of the building with tactile information

4 Scope of the Study

In 2015–2021, wayfinding systems were designed for each AFAD building, tailored to the unique character of each structure while maintaining shared features. The systems were based on consistent elements: an original set of numerals designed by the AFAD faculty and the official typography (Scala font).

For the purpose of this study, the design process was conducted in the smallest of the campus buildings. Insights from this project were then used to guide implementation across the remaining campus structures.

The adaptation project was guided by key design principles, ensuring continuity with the existing information system while enhancing accessibility. The new design extends the current system's content and integrates graphic elements such as typography, color schemes, and pictograms, all of which were already optimized for readability and contrast. Materials were selected for their visual and tactile appeal, aligning with the wayfinding system and addressing the needs of people with disabilities. The project prioritized cost-effectiveness and feasibility, ensuring that production could be carried out efficiently. Additionally, the system was designed to optimize user navigation within the building, accommodating diverse visitor needs, including those with visual impairments and blindness. The inclusive approach aimed to create a seamless, user-friendly experience for all. At the same time, emphasis was placed on usability, intuitive interaction, and minimizing cognitive effort for users navigating unfamiliar environments, following user-centered design principles (Norman, 2013).

After reviewing the accessibility audit, analyzing the existing system, and identifying its shortcomings in effectively informing all mapped user groups, the system was enhanced with the following elements: a simplified 3D tactile model of the interwar building to represent its complex layout and complementary detailed visual and tactile maps for each floor.

The spatial model provides an overview of the building's layout, including floor levels, communication pathways, the locations of key public functions, and less obvious spaces, such as workshops and restrooms on level—1 (Fig. 1).

Certain details were intentionally reduced, while others, such as the main historic staircase, were depicted with greater specificity. This approach enables users to grasp the unconventional layout of the building and develop a mental map of its spaces, using distinctive points as a reference. Such spatial representation supports cognitive mapping processes that are essential for orientation in complex architectural environments (Lynch, 1960).

Other elements of the model, such as elevators and secondary staircases, were simplified to avoid drawing undue attention, as they represent standard building features. A key decision was to mark the layout of hallways using grooves on the floor plan and provide descriptions of essential functions using raised capital letters, 1 cm in height, alongside Braille inscriptions. High-contrast color combinations were also employed

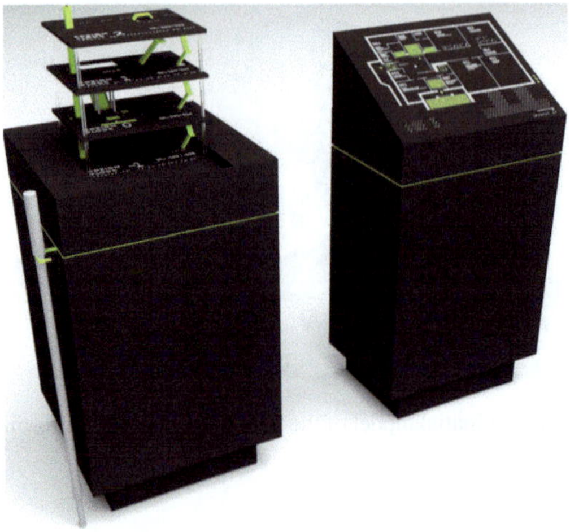

Fig. 2 Spatial model of the building and tactile map

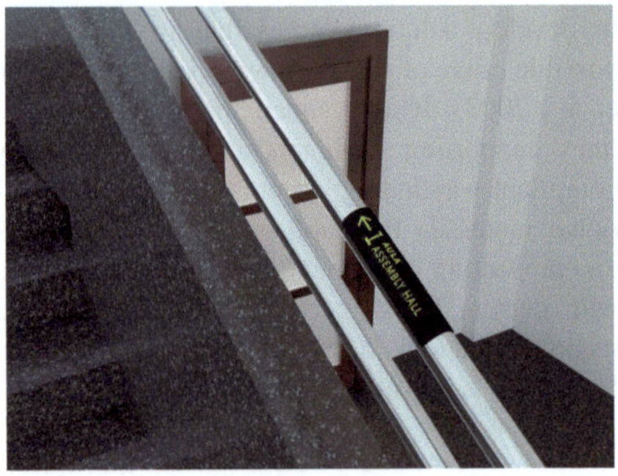

Fig. 3 Tactile handrail

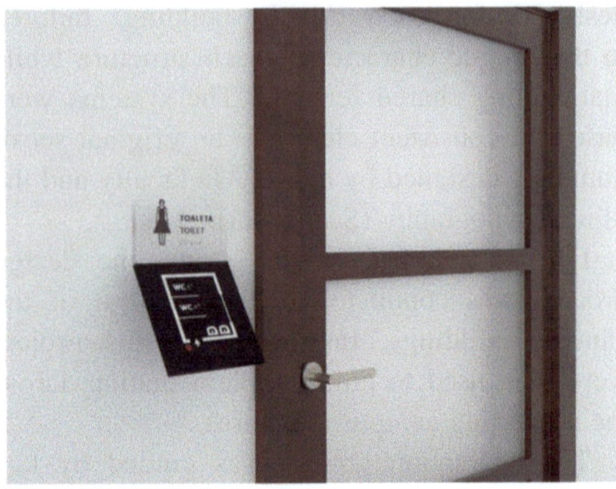

Fig. 4 Restroom tactile sign

to ensure the information is legible for sighted, visually impaired, and blind users alike.

The tactile and graphic maps display the building's full range of functions. They supplement the 3D model by offering detailed information on the arrangement and naming of functions on each floor. They include distinctive elements, such as columns and staircases, allowing users to familiarize themselves with the building's functional layout (Fig. 2).

The 3D spatial model is located on the ground floor near the reception desk. It is designed for use by both blind and sighted individuals. The reception staff can use the model to guide visitors to less obvious areas. The 2D maps are placed on every floor near the elevators. The elements complement each other, avoiding redundancy. Graphic elements ensure consistency, incorporating high-contrast colors, pictograms, typefaces, and materials.

All textual elements in the system were designed to be accessible both visually and by touch, with appropriate font sizes, strong contrast, and raised lettering. The same principles guided the design of other graphic elements, ensuring that differences between features—such as room outlines, functional designations, and hallway markings—are clearly indicated through the use of color, varying elevations, and line thickness. The implementation also followed recommendations concerning tactile information systems and orientation solutions for blind and visually impaired users in public buildings (Polski Związek Niewidomych: Projektowanie i adaptacja przestrzeni publicznej do potrzeb osób niewidomych i słabowidzących, n.d.).

The system was further enhanced with tactile signage on handrails, restroom and room markers with tactile information, tactile pathways on floors, and markings on stair treads. (Figs. 3 and 4) The existing restroom signage was supplemented with a map illustrating the layout of restroom facilities. This map is particularly valuable for persons with visual impairments: it supports their independence and helps them plan their visit without assistance.

Apart from the graphic elements, the design process included ergonomic carriers for these elements, enabling users to comfortably put their aids aside while reading tactile information The materials and style of these elements harmonize with the building's interior and existing information system.

The project employed a mix of low-prototyping techniques and cost-effective technologies, along with the integration of ready-made solutions available on the market, such as adhesive tactile dots for staircase landings. After prototyping, the solutions were tested with a small group of users. Two groups were invited to this phase: students from another university who visited the building for the first time and a blind participant accompanied by an assistant. The iterative testing process was informed by co-creation methodologies that emphasize direct engagement with users during the development and evaluation of design solutions (Sanders & Stappers, 2008).

User testing involved observation and a think-aloud protocol to capture the participants' experiences navigating a designated route. The route was designed to include elements of both the new and existing systems, encompassing visual and tactile components. The goal was to verify whether the two systems created continuity in providing information and whether users with diverse cognitive abilities could effectively rely on all the system components.

5 Conclusions

The research revealed key insights into user experiences. Sighted users preferred maps over the 3D model, as they were more familiar with this format. Initially, they found the model too complex and fragile but later acknowledged its value in understanding the building's topography. They used both new and existing system elements, perceiving the system as cohesive, though some had difficulty interpreting certain map features, such as restroom locations.

The blind participant found the spatial model intuitive, easily grasping the building's structure, pathways, and functional layout. The material was pleasant to touch, and tactile elements—raised and recessed markings, embossed letters, and Braille—were clear and unambiguous. Floor-level maps were simple to interpret, and restroom markings were accessible and well-received. However, some adjustments were needed in the handrail-based information to improve usability.

6 Overall Findings

When introducing accessibility elements, it's crucial to maintain consistency with the existing visual information system by using the same typography, colors, and graphic style. This prevents solutions from appearing "tacked on" and avoids stigmatizing users with disabilities. A combination of visual, tactile, and auditory communication ensures effective navigation for all users. Integrating graphic maps, 3D models, and Braille signage enhances accessibility. Testing with diverse users helps refine clarity and usability. Tactile information must be functional and ergonomic, with raised letters and symbols positioned for easy reading. Features like high-contrast colors, intuitive rail markings, and detailed room descriptions improve navigation for all visitors. Maintaining consistency between architectural space, graphic language, and information hierarchy is essential for the readability and effectiveness of wayfinding systems (Mollerup, 2013; Smitshuijzen, 2007). Inclusive wayfinding systems designed for users with specific accessibility needs often improve usability and spatial comprehension for broader user groups as well, confirming observations made in inclusive design research (Clarkson et al., 2003; Norman, 2013).

References

Arthur, P., & Passini, R. (1992). *Wayfinding: People, signs, and architecture*. McGraw-Hill.

Clarkson, J., Coleman, R., Keates, S., & Lebbon, C. (Eds.). (2003). *Inclusive design: Design for the Whole Population*. Springer.

Kalbarczyk, M. (n.d.). *Miasto otwarte dla niewidomych. Fundacja Szansa dla Niewidomych.* http://pliki.altix.pl/Fundacja/Miasto.pdf.. Accessed 15 Jan 2025

Kowalski, K. (n.d.). *Włącznik—projektowanie bez barier. Fundacja Integracja.* https://integracja.org/wp-content/uploads/2024/03/wlacznik_2_0_dostepny-.pdf. Accessed 15 Jan 2025.

Lynch, K. (1960). *The image of the City.* MIT Press.

Mollerup, P. (2013). *Wayshowing > Wayfinding: Basic and interactive.* BIS Publishers.

Norman, D. A. (2013). *The Design of Everyday Things.* Revised and Expanded Edition. Basic Books.

Państwowy Fundusz Rehabilitacji Osób Niepełnosprawnych: Raport końcowy z badania potrzeb osób niepełnosprawnych w Polsce. 26. (n.d.). https://www.pfron.org.pl/instytucje/badania-i-analizy-naukowe/raport-koncowy-z-badania-potrzeb-osob-niepelnosprawnych-w-polsce-2024-r. Accessed 12 May 20.

Polski Związek Niewidomych: Projektowanie i adaptacja przestrzeni publicznej do potrzeb osób niewidomych i słabowidzących. (n.d.). https://pzn.org.pl/wp-content/uploads/2024/06/Projektowanie-i-adaptacja-przestrzeni-do-potrzebosob-niewidomych-i-slabowidzacych-pelna-wersja-1.pdf. Accessed 12 May 2026.

Sanders, E. B.-N., & Stappers, P. J. (2008). Co-creation and the new landscapes of design. *CoDesign, 4*(1), 5–18.

Smitshuijzen, E. (2007). *Signage design manual.* Lars Müller Publishers.

Urząd Miasta Warszawy: Standardy zapewniania dostępności. (n.d.). https://wsparcie.um.warszawa.pl/standardy-dostepnosci. Accessed 12 May 2026.

Zimmerman, J., Forlizzi, J., & Evenson, S. (2007). Research through design as a method for interaction design research in HCI. In *Proceedings of the SIGCHI conference on human factors in computing systems* (pp. 493–502). ACM.

O'Culo, A Serious Game on Culture-Sensitive Design

Kristof Vaes and Doris Van Boxem

Abstract

O'CULO is a serious game developed to enhance cultural awareness and sensitivity among design students and professionals. This paper details the development, gameplay, and impact of O'CULO, emphasizing its role in fostering cultural dialogue and introspection within design practices. The game addresses the necessity of interculturalization in design, driven by globalization and superdiversity, and underscores that cultures are not homogeneous entities. The iterative development process of O'CULO involved extensive literature research, psychological frameworks, and expert contributions, resulting in a game that promotes cultural sensitivity through engaging activities. O'CULO is structured into three levels: Cultural Encounters, Culversations, and Culture Cross, each targeting different aspects of cultural sensitivity. The game includes activities such as quizzes, discussions, and scenario analyses, which encourage players to confront and reflect on their cultural biases. The final level, Product Prophecy, presents design mismatches due to cultural differences, offering lessons on the importance of cultural awareness in design. The game concludes with a quartet game that highlights successful culture-aware designs and provides practical tips for real-world application. Tested in Inclusive Design courses and workshops, O'CULO has demonstrated its effectiveness in promoting cultural awareness and dialogue. Feedback from students and professionals indicates increased recognition and respect for cultural diversity. The game's structure, real-world examples, and practical tips enhance its educational value, making it a valuable tool for fostering cultural sensitivity in design. Future research could explore the long-term impact of O'CULO on design practices and its adaptation for various educational and professional contexts. Digital versions of the game could further expand its accessibility and interactive features, ensuring its continued relevance and effectiveness.

Keywords

Design for diversity · Inclusive design · Culture-sensitive design · Serious game

1 Introduction

In our diverse world, designers must consider mixed cultural contexts beyond their own to avoid 'mismatches' and product rejection (Geldof, 2019; van Boeijen & Stappers, 2011; Hao, 2019). Products and services convey diverse meanings across cultures; designers must

K. Vaes (✉) · D. Van Boxem
Faculty of Design Sciences, Product Development, University of Antwerp, Antwerp, Belgium
e-mail: kristof.vaes@uantwerpen.be

S. Camplone and C. Porfirione (eds.), *Design for All and the Future of the Planet*, Design For Inclusion 6, https://doi.org/10.1007/978-3-032-21769-1_16

acknowledge their cultural biases, shaped by their cultural background, customs, and values. O'CULO, a serious game developed by Doris Van Boxem and Kristof Vaes at the University of Antwerp, addresses this need by encouraging players to engage with cultural differences and understand the impact of cultural contexts on design. The name "O'CULO" is a play on words, combining "ocular" (relating to lenses) and "culture," but also humorously noting its slang meaning in Spanish, which underscores the importance of cultural awareness.

2 Development of O'CULO

The development of O'CULO was an iterative process involving literature research, psychological frameworks, and expert contributions. The game went through several prototypes and co-creation sessions with students and experts to refine its content and gameplay (Fig. 1).

- **Literature Research**: The foundation of O'CULO is built on extensive research into cultural sensitivity and design. Key references include works by Geldof (2019), van Boeijen and Stappers (2011), and Hao (2019), which emphasize the importance of acknowledging cultural biases and the diverse meanings products and services convey across cultures.
- **Psychological Frameworks**: In developing the psychological framework, the initial focus was on identifying the appropriate game format and then translating it into a safe and educational environment. A serious game was chosen for its ability to engage players and facilitate reflection and feedback (Laamarti et al., 2014; Lameras et al., 2017). The Triadic Game Design (TGD) approach provided a general design philosophy (Harteveld, 2011). Emphasis was placed on creating a psychologically and emotionally safe environment (Santl, 2023). Expert interviews with psychologists helped identify potential risks and opportunities. Santl (2023) highlights the importance of validating players' feelings and fostering dialogue. To ensure everyone feels safe and comfortable, all players follow the 'Protocul', creating an atmosphere of openness and appreciation. This 'Protocul' outlines the game's code of conduct to ensure safety, objectivity and neutrality.
- **Expert Contributions**: Input from design professionals and cultural experts ensures the game's relevance and accuracy. These contributions help tailor the game's content to address real-world cultural challenges in design.

The iterative process involved creating and customizing several research prototypes, with a total of four co-creation sessions and game sessions with design students and experts. This approach ensured that the game was continuously refined based on feedback and practical insights.

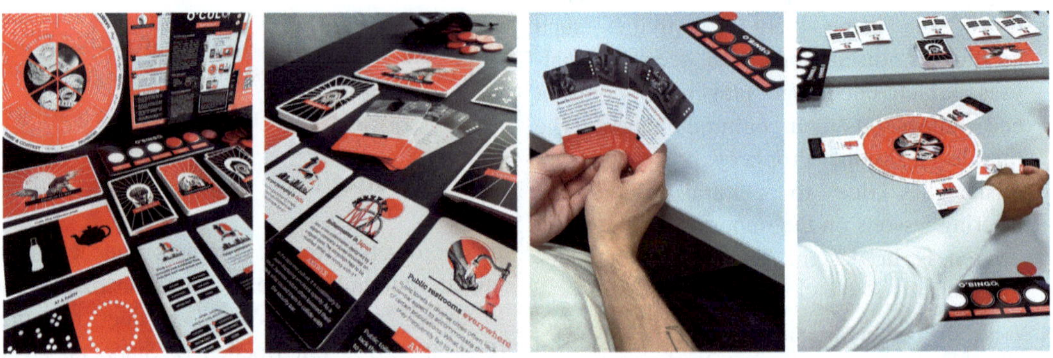

Fig. 1 Overview of the physical elements of the O'Culo serious game and images of the gameplay during the workshops

3 Gameplay and Structure

O'CULO is structured into three levels, each designed to address different aspects of cultural sensitivity in design. The game is intended for 4 to 10 players and takes approximately 2 hours to complete.

3.1 Cultural Encounters

The first level introduces players to cultural differences and challenges them to confront cultural biases (Fig. 2).

- **Culversations**: In this mini-game, a player states a belief about their own culture that they think doesn't apply to the other players. Others raise their hands if it does apply to them or cross their arms if it doesn't. The group then discusses their answers. This activity encourages players to share and reflect on their cultural backgrounds, including family, hobbies, friends, and food.
- **Culture Cross**: This activity involves a quiz question related to cultural knowledge. Players discuss their answers in a group and then reveal the correct answer from the quiz card. This activity promotes group discussion and learning about different cultures.
- **To Me or to We?**: These cards show an illustration depicting two ways to handle a particular situation. The group discusses whether the behavior of a person in this situation is determined individually or culturally. This activity helps players understand the influence of culture on behavior.

3.2 Product Prophecy

In this level, players are presented with intriguing examples of design mismatches, where products or services fail to resonate with consumers due to cultural differences. Players take turns reading a card describing a design mismatch. The others guess what the mismatch was or its cause, with the reader providing hints. Once the group guesses correctly, they categorize it on the Wheel of Categories. This level provides valuable lessons on the impact of poor cultural awareness in design (Fig. 3).

3.3 For Keepsake!

In the final level, players dive into a quartet game, collecting cards featuring inspiring examples of successful culture-aware design. This level offers practical tips for applying these principles in real-world scenarios, empowering players to create designs that resonate across cultures. Players learn about best practices in culturally sensitive design and how to implement these practices in their work.

The game also includes a set of guidelines, referred to as the 'Protocul', to ensure

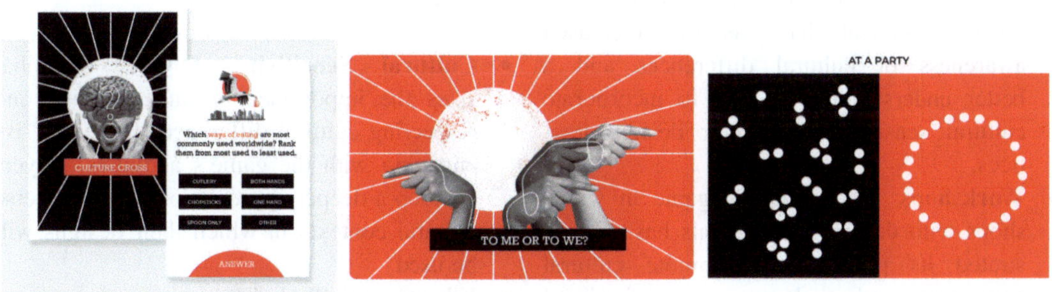

Fig. 2 O'Culo—level 1: Front and back of Culture Cross card (left) and To me or to We? Card

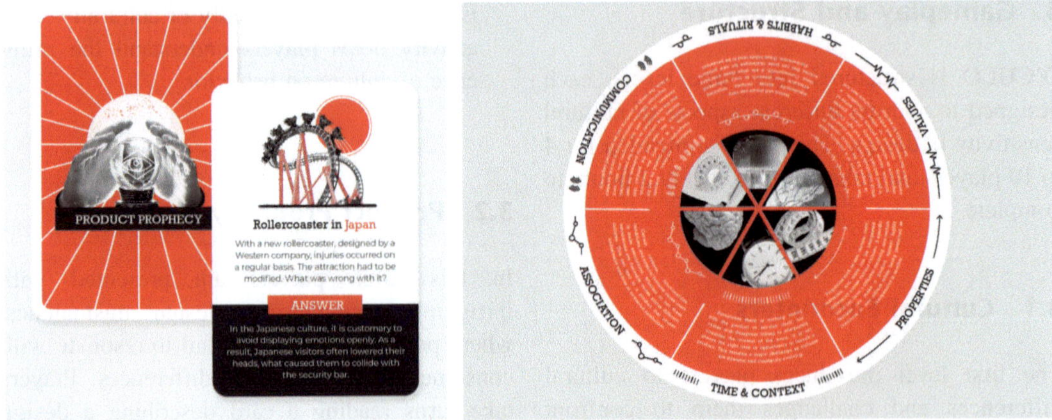

Fig. 3 O'Culo—level 2: Front and back of Product Prophecy card (left) and the Wheel of categories (right)

a positive and productive gaming experience. These guidelines emphasize the importance of open-mindedness, the acceptance of mistakes, and focusing on problem-solving rather than team competition. The Protocul helps create a safe environment where players feel comfortable expressing their thoughts and learning from each other.

4 Impact and Validation

O'CULO has been tested in Inclusive Design courses at the Faculty of Design Sciences and evaluated in a E&PDE workshop in Barcelona (September 2023).

- **Inclusive Design Courses**: Testing the game in real educational settings has provided valuable feedback and insights into its effectiveness. Students have reported increased awareness of cultural differences and a better understanding of how to incorporate cultural sensitivity into their design practices.
- **Workshops**: Evaluating the game in workshops with design professionals has demonstrated its potential to promote cultural awareness and dialogue in professional settings. Participants have appreciated the game's interactive and engaging approach to learning about cultural sensitivity.

The results obtained from these tests and evaluations indicate that O'CULO not only promotes awareness of cultural diversity in design but also positively impacts students from migrant backgrounds. The game fosters recognition, awareness, and respect, contributing to a more inclusive and culturally sensitive design practice (Fig. 4).

5 Conclusions

O'CULO successfully addresses the need for cultural sensitivity in design by providing an engaging and educational platform for players to explore cultural differences. The game's iterative development process and validation through testing have ensured its effectiveness in fostering cultural awareness and dialogue.

- **Cultural Sensitivity**: O'CULO emphasizes the importance of understanding and respecting cultural differences in design. By engaging with the game's activities, players develop a deeper appreciation for the diverse cultural contexts in which their designs will be used.
- **Educational Tool**: The game serves as an educational tool to promote cultural awareness among design students and professionals. Its interactive and hands-on approach

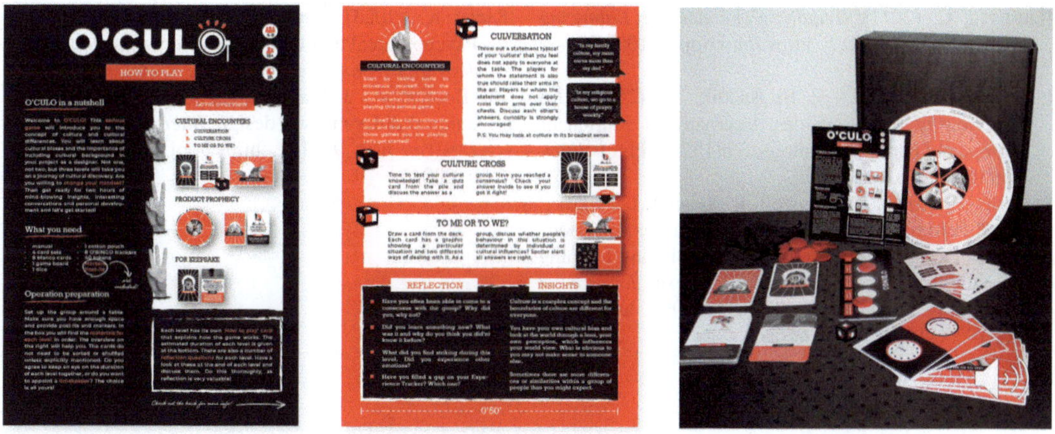

Fig. 4 O'Culo 'How to Play' instruction card (left) and overview of all elements of the O'Culo serious game (right)

makes learning about cultural sensitivity both enjoyable and impactful.

- **Iterative Development**: The continuous feedback and improvement through co-creation sessions have ensured that O'CULO remains relevant and effective. The game's development process highlights the importance of involving diverse perspectives in creating educational tools.

The game's structure, with its three levels of gameplay, provides a comprehensive approach to learning about cultural sensitivity. Each level builds on the previous one, gradually deepening players' understanding of cultural differences and their impact on design. The inclusion of real-world examples and practical tips further enhances the game's educational value.

Future research could explore the long-term impact of O'CULO on design practices and its potential adaptation for other educational contexts. Additionally, expanding the game's content to include more examples of cultural mismatches and successful culture-aware designs could further enhance its educational value.

Future directions for O'CULO could also involve adapting the game for different educational levels and professional contexts. For example, the game could be tailored for use in primary and secondary education to introduce younger students to the concept of cultural sensitivity. Similarly, it could be adapted for use in corporate training programs to promote cultural awareness among professionals in various industries.

Another potential direction is the development of digital versions of O'CULO, which could make the game more accessible to a wider audience. Digital platforms could offer interactive features, such as virtual co-creation sessions and online discussions, to enhance the learning experience. Additionally, digital versions could include multimedia elements, such as videos and animations, to illustrate cultural concepts and design examples more vividly.

Finally, ongoing research and feedback from users will be crucial in continuously improving O'CULO. Regular updates to the game's content and structure, based on the latest research and user experiences, will ensure that it remains a relevant and effective tool for promoting cultural sensitivity in design.

References

Geldof, M. (2019). Cultural sensitivity in design. *Journal of Design Research, 17*(2), 123–145. https://doi.org/10.1504/JDR.2019.10012345

Hao, Q. (2019). Intercultural design: Bridging cultural gaps. *Design Studies, 40*(3), 201–218. https://doi.org/10.1016/j.destud.2019.03.001

Harteveld, C. (2011). *Triadic game design: Balancing reality*. Springer. https://doi.org/10.1007/978-1-84996-157-8

Laamarti, F., Eid, M., & El Saddik, A. (2014). An overview of serious games. International Journal of Computer Games Technology, 358152. doi:https://doi.org/10.1155/2014/358152.

Lameras, P., Arnab, S., Dunwell, I., Stewart, C., Clarke, S., & Petridis, P. (2017). Essential features of serious games Design in Higher Education: Linking learning attributes to game mechanics. *British Journal of Educational Technology, 48*(4), 972–994. https://doi.org/10.1111/bjet.12467

Santl, K. (2023). Creating safe game environments: Psychological and emotional safety in serious games. *Journal of Game Design, 15*(2), 45–60.

van Boeijen, A., & Stappers, P. J. (2011). Understanding cultural differences in design. *International Journal of Design, 5*(2), 99–112. https://doi.org/10.1162/IJOD_a_00045

Integrating Design for All in Bathroom Furnishings: Prime and Shine Washbasins

Francesco Rodighiero

Abstract

The Prime washbasin introduces a new design experience where functionality for everyone goes beyond the usual formal trends of countertop washbasins. This is something absolutely innovative in the world of contemporary bathroom furnishings, never before considered for such a broad user base. And it applies not only to domestic environments, but also to public restrooms and hospitality spaces, where compliance with regulations is not optional.

Keywords

Washbasin · Accessibility · Elegance · Functionality

The Prime and Shine projects embody a groundbreaking vision of inclusive design, developed through a co-design process with users. Prime, in particular, stands out as the first countertop washbasin specifically designed for individuals with disabilities who retain a degree of autonomy, offering an ideal solution for hospitality settings. Its innovation lies in the ability to seamlessly combine accessibility, elegance, and functionality in a single product, without compromise.

Shine, the natural evolution of Prime, expands on its formal and functional language, reinforcing the message that accessibility is not merely a technical requirement but a cultural value that can be integrated with style and modernity.

A pivotal element in the success of Prime has been the recognition of its innovation through television media. Despite the use of traditional and digital communication tools, it was the visibility achieved through Italian television series such as *I Fantastici 5 and Lea* that brought the true value of the project to light. Television enabled a broader audience to understand, appreciate, and place trust in Prime, underscoring how it remains, even today, a vital medium for advancing the culture of inclusive design, transcending the limitations of social networks and other conventional media.

F. Rodighiero (✉)
Milan, Italy
e-mail: francesco@rodighiero.design

S. Camplone and C. Porfirione (eds.), *Design for All and the Future of the Planet*, Design For Inclusion 6, https://doi.org/10.1007/978-3-032-21769-1_17

Afterword

The Gaia Declaration Ceremony

Giuseppe Di Bucchianico

The international conference "*Gaia 2024. Design for All and the Future of the Planet*", held in Vila Nova de Gaia on 6th and 7th June 2024, was not only notable for the high-level scientific content presented during the sessions and panels, nor for the thought-provoking reflections offered by the keynote speakers—each of whom contributed with institutional depth and clarity—but also represented a highly meaningful moment in the history of the international Design for All community that gathers around EIDD—Design for All Europe.

At the close of the second day, on the afternoon of June 7th, a ceremony was held in a particularly symbolic location for the city of Vila Nova de Gaia: the Corpus Christi Convent. This event will likely stand as a historic milestone for Design for All in the coming decades.

In fact, in this meaningful setting, the association EIDD—Design for All Europe, in the presence of various local authorities, officially presented the *Gaia Declaration*, a new milestone in the ongoing path toward a more inclusive and equitable society.

To fully understand the significance of the *Gaia Declaration*, it is necessary to take a step back.

EIDD—Design for All Europe was founded in 1993. More than three decades have passed since its founding, and twenty years since the drafting of the *Stockholm Declaration*, a document that, thanks to the collective efforts of the community that identified with the principles of Design for All at the time, provided the first definition of Design for All as "design for human diversity, social inclusion and equality." This definition has since been embraced by leading international authorities and organizations.

In 2023, during the EIDD General Assembly in Bratislava, which also marked the 30th anniversary of the association, it was agreed that the time had come to revisit the principles of the *Stockholm Declaration* in light of the urgent new global challenges posed by phenomena such as climate change, widespread conflict, ageing populations, and the unchecked acceleration of technological development.

Following a full year of collaborative drafting and careful review of the concepts, the terminology and every single word, with wide and inclusive participation from the EIDD Member Organizations, the *Gaia Declaration* was finally presented on June 7th, 2024, in a touching and formal ceremony at the historic Corpus Christi Convent in Vila Nova de Gaia, Portugal.

S. Camplone and C. Porfirione (eds.), *Design for All and the Future of the Planet*, Design For Inclusion 6, https://doi.org/10.1007/978-3-032-21769-1

From that day forward, a process of endorsing the document was launched. It immediately attracted dozens of in-person signatories and has since continued online, collecting further endorsements from key figures in politics, culture, economics, and civil society, most importantly, from individual citizens who identify with and support the values and principles expressed in the Declaration.

The *Gaia Declaration* is intentionally concise: like the *Stockholm Declaration*, it fits on a single page, so it can be read in its entirety within a few minutes, although the ideas it conveys are far-reaching and invite deep reflection.

The text is structured into two main parts. It is thus a document that is accessible yet rich in meaning and values, and its name reflects a felicitous combination: it evokes both the host city of the conference, *Vila Nova de Gaia*, and the ultimate addressee of the Declaration, planet Earth itself, whose ancient name is *Gaia*.

As for the content and structure of the document, it opens with a Preamble that reconnects with the *Stockholm Declaration* of 2004 and highlights some of the pressing phenomena affecting contemporary societies–and the planet at large–that threaten the core principles of inclusion, diversity, and equality upon which Design for All is founded.

The second part of the document, building on this analysis, proclaims a series of principles that the international Design for All community is called to embrace. These principles are presented as challenges to be addressed in the coming decades, with a strategic vision. The *Gaia Declaration* calls special attention to the following phenomena:

Climate Change

This is now an undeniable and scientifically proven reality, generating not only environmental consequences but also acting as a multiplier of humanitarian emergencies on a global scale. For this reason, the *Gaia Declaration* calls upon all those operating within a Design for All framework to take into account the environmental impacts of their actions and design choices, recognizing that human diversity is itself an essential part of the richness and complexity of the Earth's ecosystem. Humanity has an ancestral bond with Nature, rooted in its evolutionary history, where humans and the environment continuously influence each other in a systemic relationship.

Ageing Populations

Although the widespread ageing of humanity is one of the greatest achievements in the history of human progress, it also presents a major challenge, one that must be addressed first by policymakers, but also by designers and planners. A radical and innovative shift in how we conceive of living environments is needed, with the awareness that the elderly condition, with its strengths but also its limits and vulnerabilities, will become increasingly predominant in the future of humanity. Moreover, ageing will represent a strategic and decisive factor in future choices related to economic policy and welfare systems.

Wars and Social Conflicts

While conflicts have always existed, they now tend to intensify within a globalized society, one that is economically interconnected, culturally diverse, and increasingly polarized between opposing living conditions and economic opportunities. The growing economic, cultural, and social gaps between population groups, coupled with widespread insecurity and distrust of diversity, create tensions that often escalate into violent conflict and humanitarian emergencies. A multilevel response is needed: one that addresses immediate crises in the short term, while fostering long-term dialogue and reconciliation, and emphasizing the value of diversity.

Exponential Development of New Technologies

Digital technologies have become pervasive in contemporary societies. They offer extraordinary opportunities that were unimaginable just a few years ago, with their potential expanding at an exponential pace. However, they also give rise to new forms of social alienation, especially among generations that struggle to keep up with technological evolution. In this context, new technologies must take on a vital challenge: to engage meaningfully with the skills and needs of multiple generations, potentially far apart, and to serve as tools of inclusion, rather than means of exclusion, for those who face greater difficulty and are at risk of being left behind.

These phenomena represent the new challenges that humanity must face, both now and in the near future. Through the *Gaia Declaration*, addressed to political, economic, professional, and technical decision-makers, a call is made for swift and coordinated action, grounded in the principles and participatory, conscious approach of Design for All.

The following pictures (Figs. 1, 2, 3) were taken during the conference. They show EIDD members and conference participants. The pictures were taken by the organizers, who own copyright.

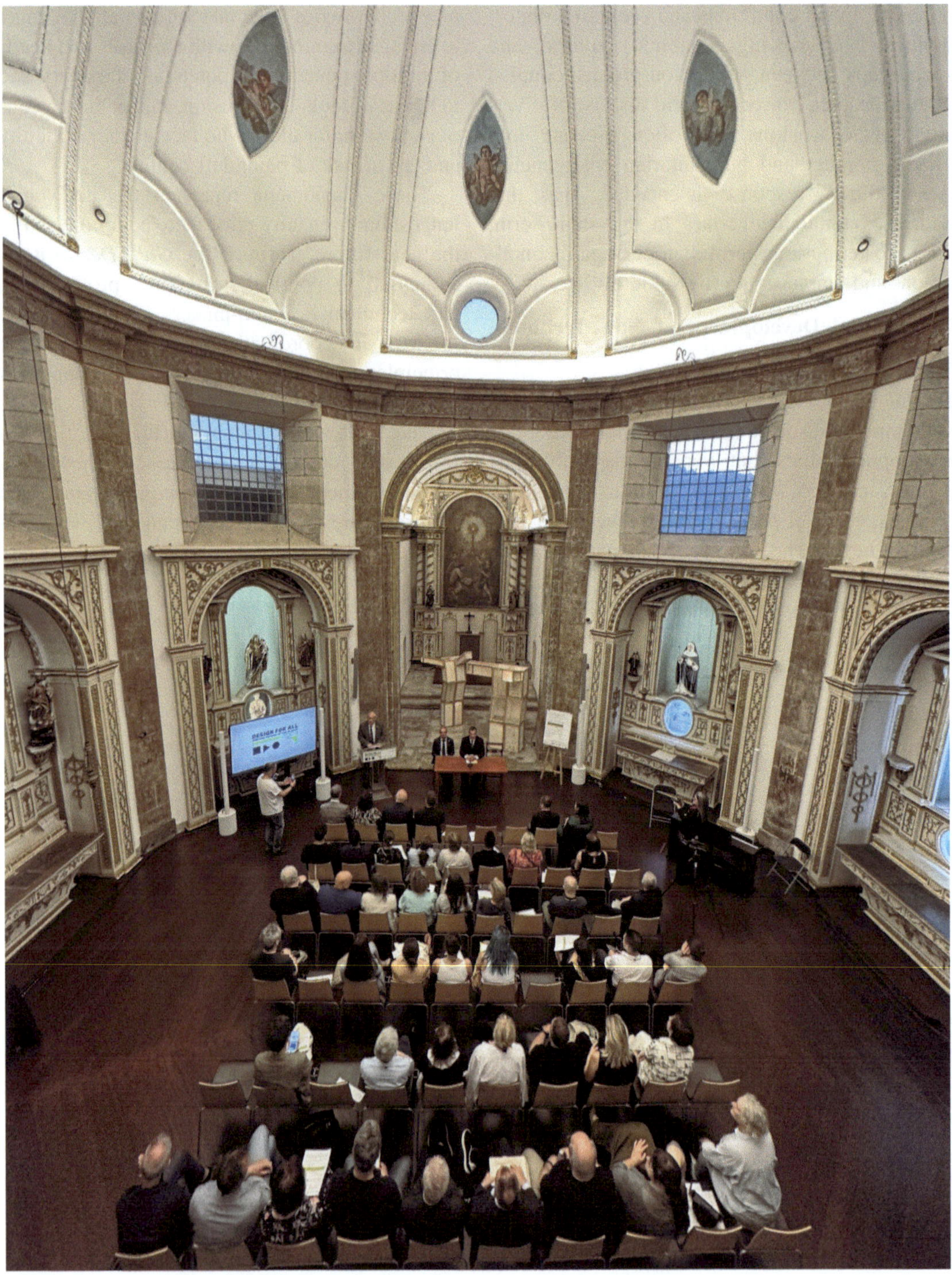

Fig. 1 Participants listening to one of the conference lectures at GAIA 2024

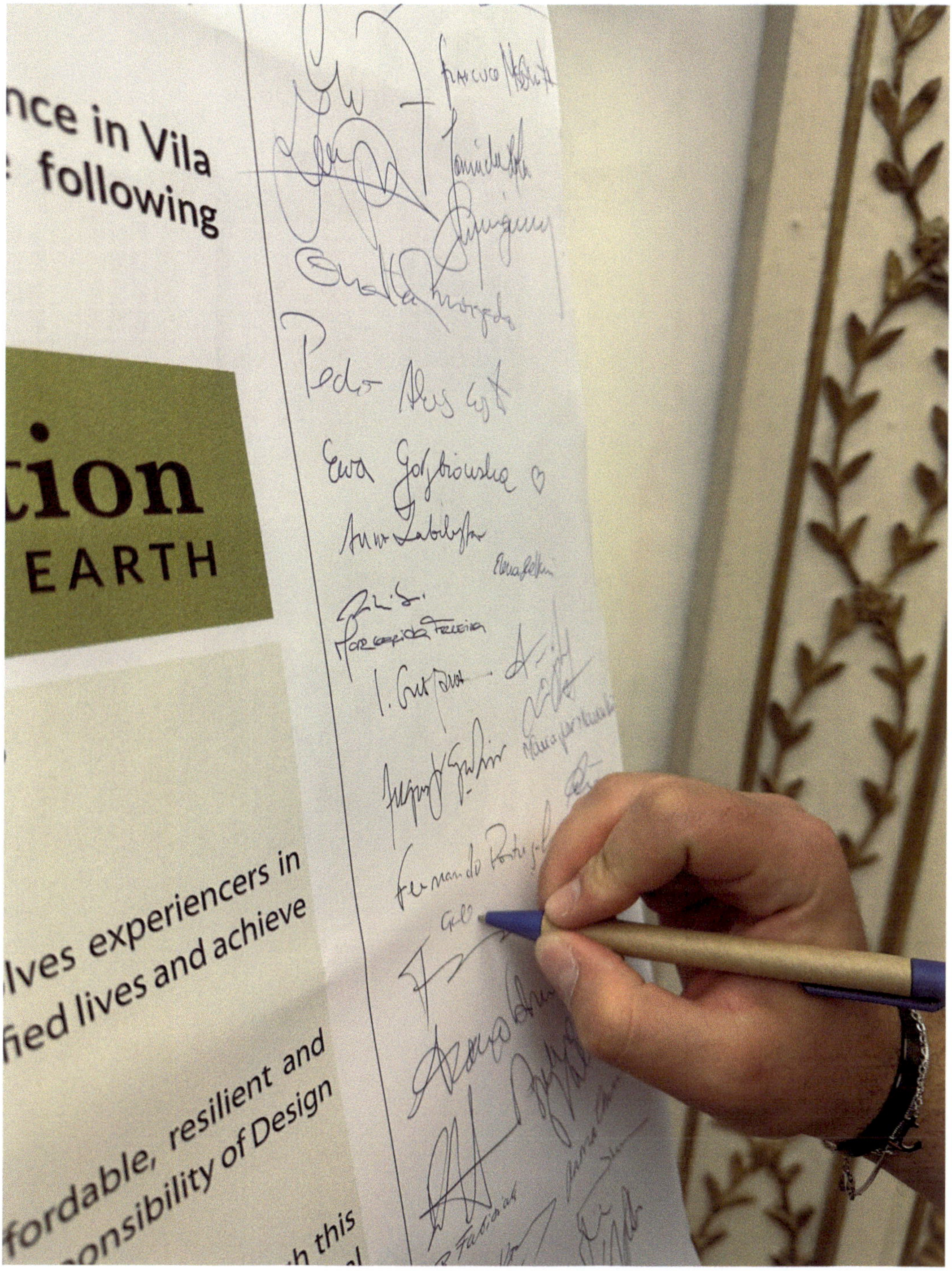

Fig. 2 One of the participants is signing the GAIA declaration at GAIA 2024

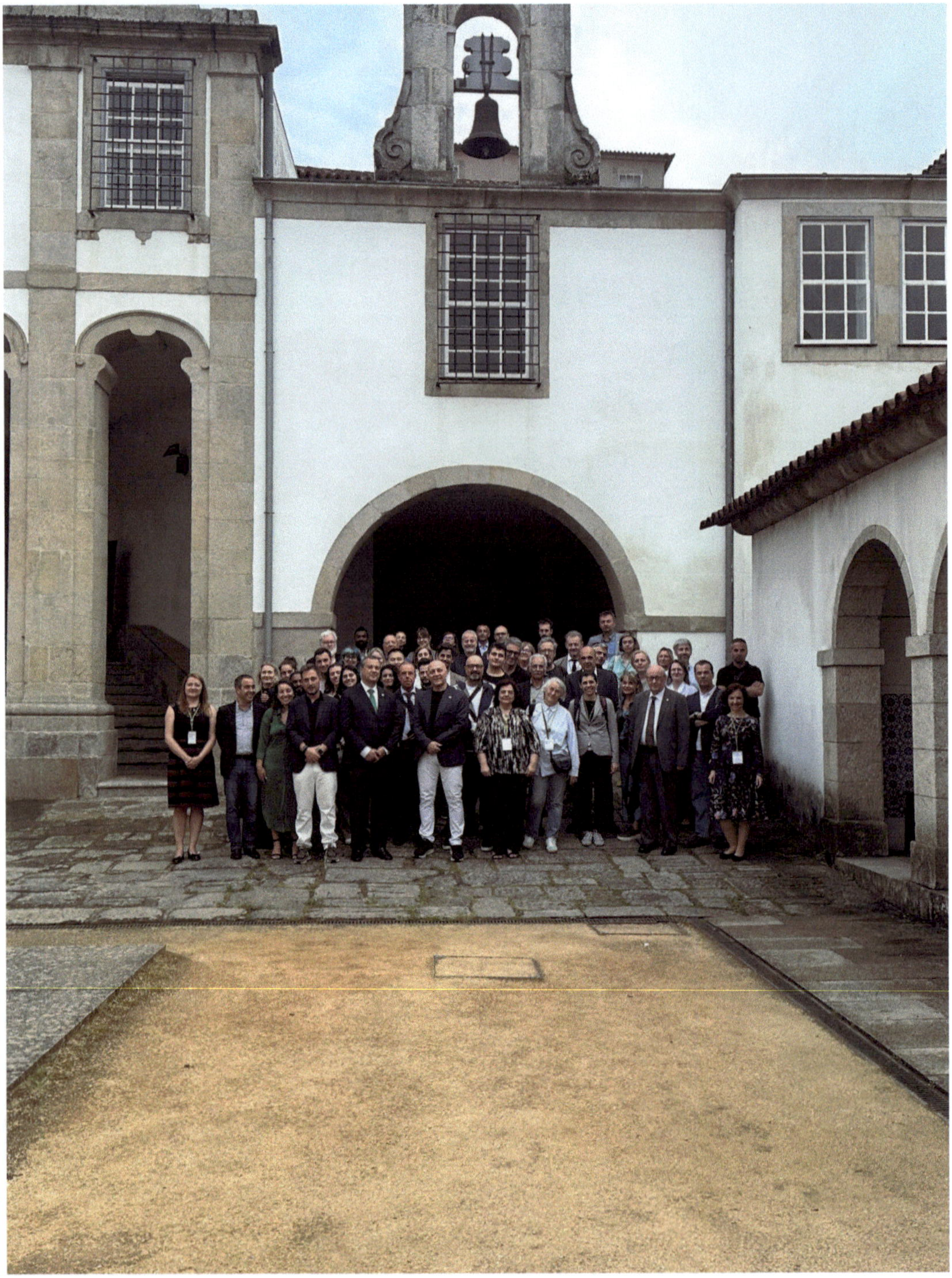

Fig. 3 Group photograph taken during GAIA 2024

The EIDD Gaia Declaration (Draft)

The Design for All Statement for the Earth

"Good design enables, bad design disables" (Paul Hogan, Dublin, 3 April 1993)

Preamble

Established in 1993, EIDD—Design for All Europe has kept faith with the mission statement: "Enhancing the quality of life through Design for All". In 2003, it started working on the definitive description of the origins, philosophy and methodology of Design for All: the EIDD Stockholm Declaration, adopted in Stockholm on 9 May 2004, including the essential definition: **Design for All is design for human diversity, social inclusion and equality**. This firmly established Design for All as the most effective methodology, in the mainstream of academia and professional practice, for the social inclusion of human diversity in today's rapidly-changing world.

Now, climate change, widespread warfare, the ageing population and the runaway development of technology present new and pressing challenges, including migration, natural disasters, the refugee crisis and alienation, as phenomena of potential exclusion that call for innovative responses from the design community worldwide.

At its 2023 General Assembly in Bratislava, EIDD—Design for All Europe resolved to create a new document to express how Design for All can contribute to these developments.

EIDD—Design for All Europe, on the occasion of its Annual Conference in Vila Nova de Gaia, Portugal, on 7 June 2024, therefore presents the following Declaration:

Climate change is an incontrovertible fact. In this new context, Design for All involves experiencers in devising a constant flow of diverse viable responses to enable all people to live dignified lives and achieve their highest potential and aspirations.

A sustainable agenda can only be complete if it is environmentally friendly, affordable, resilient and socially aware, i.e. truly useable by everyone in ever-changing societies. This is the responsibility of Design for All.

Our societies are culturally, ethnically and religiously diverse and most are ageing rapidly. Although this growing diversity is evidence of the most successful advance in human history, it is a challenging political reality that calls for a radical shift in thinking that can only come from the practice of Design for All: continuous consultation, education and awareness-raising for inclusion.

Violence, strife, conflict and emergencies are always present. Design for All responds with its inclusive, experiencer-centred process, generating in the short term emergency responses; in the medium term reconstruction; in the long term the peace process itself.

Digital worlds also matter: Artificial Intelligence and other emerging technologies have an

S. Camplone and C. Porfirione (eds.), *Design for All and the Future of the Planet*, Design For Inclusion 6, https://doi.org/10.1007/978-3-032-21769-1

important role to play as partners with Design for All to achieve more inclusion.

EIDD—Design for All Europe therefore calls on the European and international institutions, national, regional and local governments and professionals, businesses and social actors to take all appropriate measures to implement Design for All in their policies and actions. Time is running out: action is needed now, in the name of Gaia, in the name of the Earth itself.

MIX
Papier aus verantwortungsvollen Quellen
Paper from responsible sources
FSC® C105338

If you have any concerns about our products,
you can contact us on
ProductSafety@springernature.com

In case Publisher is established outside the EU,
the EU authorized representative is:
Springer Nature Customer Service Center GmbH
Europaplatz 3, 69115 Heidelberg, Germany

Printed by Libri Plureos GmbH
in Hamburg, Germany